THE
CURIOUS
WORLD OF
BACTERIA

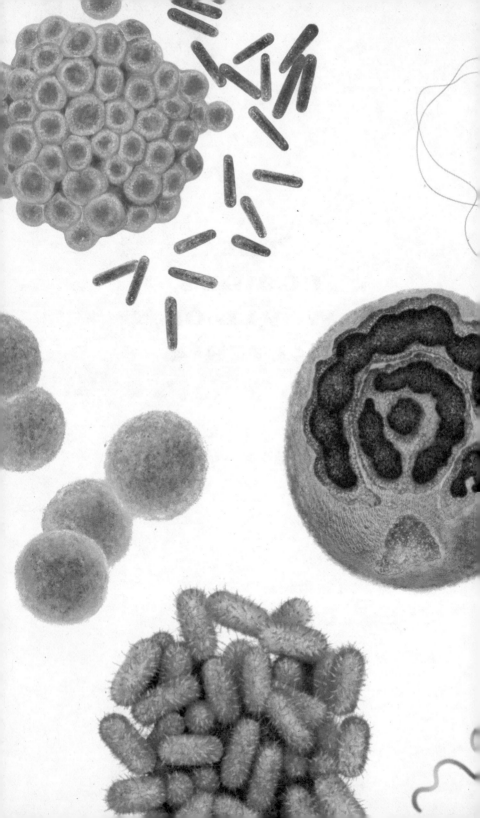

THE CURIOUS
WORLD OF
Bacteria

Ludger Wess

Translated by Jamie McIntosh
Illustrations by Falk Nordmann

GREYSTONE BOOKS
Vancouver/Berkeley/London

Greystone Books Ltd.
greystonebooks.com

Cataloguing data available from Library and Archives Canada
ISBN 978-1-77164-825-7 (cloth)
ISBN 978-1-77164-826-4 (epub)

Editing by Rowena Rae
Proofreading by Dawn Loewen
Cover and text design by Jessica Sullivan
Cover and interior illustrations by Falk Nordmann

Printed and bound in Canada on FSC® certified
paper at Friesens. The FSC® label means that materials
used for the product have been responsibly sourced.

Greystone Books thanks the Canada Council for the Arts,
the British Columbia Arts Council, the Province of British
Columbia through the Book Publishing Tax Credit, and the
Government of Canada for supporting our publishing activities.

The translation of this work was supported by a grant from the Goethe-Institut.

Greystone Books gratefully acknowledges the xʷməθkʷəy̓əm (Musqueam),
Sḵwx̱wú7mesh (Squamish), and səlilwətaɬ (Tsleil-Waututh) peoples on
whose land our Vancouver head office is located.

Contents

Preface ix
Introduction 1

Occupiers of Technological Habitats 115

Exotic Eaters 139

Helpful Bacteria 153

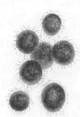

Dangerous Bacteria 183

PREFACE

IT WAS ON A VISIT to the natural history museum in Münster, Germany, when—as a seven- or eight-year-old—I first peered through a microscope. I immediately fell in love with the secret world within the water droplets on display—*Paramecium*, *Vorticella*, rotifers, microalgae, and the pounding heart of a water flea. A whole universe was revealed that had been invisible to my naked eye. What a discovery!

My enthusiasm didn't go unnoticed by my parents. They encouraged me and soon I had a junior microscope, a small aquarium, and a subscription to the scientific journal *Mikrokosmos*. I learned that this magic wasn't limited to water from puddles, ponds, and streams. The soil from potted plants, rotten apples, bits of bark, and even my own saliva were all teeming with tiny creatures.

I learned that you could culture (cultivate) microorganisms, or microbes. My pocket money disappeared on test tubes, petri dishes, and conical flasks from the local medical supply store. Unthinkable nowadays, the pharmacy in town supplied me with chemicals so that I could prepare specimens for the microscope: benzene, toluene, xylene, benzaldehyde, formaldehyde, isopropyl alcohol, and Canada balsam. I ordered scientific articles from the interlibrary loan service at our local public library. One or two weeks later, copies arrived on oversized sheets of gray Xerox paper. My room stank alternately of pungent chemicals and musty hay extracts; in the kitchen, I cooked up culture media, and I used the oven to sterilize petri dishes and large agar plates for culturing yeast and bacteria. My pride and joy was a bacteria lamp similar to the one Czech botanist Hans Molisch demonstrated to Austrian emperor Franz Joseph I in 1901. In a huge flask, I succeeded in culturing phosphorescent bacteria that I had isolated from the scales of a herring. In this wan bacterial light, I was able to read in my darkened room—a pleasure that lasted for only one night, however.

My interest in bacteria was linked to other exciting events in my childhood—the moon landing and the photos and measurements sent back from surveys of Venus and Mars. Would they find traces of life there, and maybe even some bacteria?

The initial data transmitted from Venus and Mars were of little value, but in 1971, excited NASA scientists announced the discovery of the basic components of life in a number of meteorites. As a result of some cosmic collision between two celestial bodies, would it be possible for bacteria to travel right across the universe to other planets and moons and gain a foothold there? Could there be some truth in the panspermia hypothesis—that primitive life-forms spread through the universe via debris from other worlds? Is that how life started on Earth?

I devoured articles and books on astrobiology, a new scientific discipline investigating the prerequisites for life, and was amazed by the extreme conditions that bacteria could endure on Earth.

The more I learned, the more questions surfaced, and finally, I studied biology—of course—to conduct my own research.

Even though I no longer do research myself but instead write about what others have discovered or want to discover, my interest in the invisible universe all around us has not diminished. I'm still fascinated by the sheer adaptability of bacteria that manage to find nutrition in places that, in our opinions, are empty and desolate. Above all, I've been interested for years in the extremophiles among these organisms—those living in extreme conditions that, from our human perspective, seem to be impossibly hostile to life: hot springs, icy or arid landscapes, acidic or alkaline environments, extreme pressure.

For this book, I decided to introduce bacteria that represent the vast range of life-forms and habitats occupied by these organisms. I also describe bacteria that are crucial to our daily life and economy, as well as to health and illnesses. Finally, I include bacteria that, in the coming years, will benefit people by helping to dispose of garbage and residues, saving energy, and making foods more efficient to produce and better at sustaining our health.

The result is fifty portraits—a minuscule selection—but enough, I hope, to captivate readers and arouse greater interest in the most versatile and astonishing of all living organisms.

LUDGER WESS *Gaiole in Chianti/Hamburg, 2020*

INTRODUCTION

EARTH IS GOVERNED BY BACTERIA. A small amount of soil—less than one-tenth of a cubic inch, or one cubic centimeter—contains roughly a billion of them. A teaspoonful of pond water has about a million, and even a small volume of air—say, 35 cubic feet, or one cubic meter—has a good thousand. Animals come and go, but bacteria stay, and so far, they have survived every catastrophe in Earth's history.

Bacteria are thought to have originated at least 3.8 billion years ago, probably on the ocean floor near hydrothermal vents spewing hot water with a high mineral content. Even today, the majority of all bacteria on the planet live in the sea or beneath the seabed. This is hardly surprising, as two-thirds of the world's surface is covered by water with an average depth of almost 2.5 miles (4 kilometers), so 99 percent of our planet's biosphere consists of salt water, along with salty sediments and soils.

Only recently were bacteria discovered at a depth of 3 miles (5 kilometers) below the Earth's crust. The total subterranean biosphere probably harbors 16.5 to 25 billion tons (15 to 23 billion metric tons) of bacteria—that's one hundred times as much as the total weight of the world's entire human population. If you were to take all the genetic material of all bacteria living on our planet and place it end to end, it would stretch to the borders of the observable universe.

Starting from the oceans, bacteria have spread over the entire planet. We can find them today in extremely hot, cold, dry, and humid regions. They can be deep inside rocks, on Himalayan peaks, in salt crystals, and in acids and alkalies. They can live alongside heavy metals, in the innermost parts of atomic reactors, and in volcanic glass. Even deserts, with less than 10 inches (25 centimeters) of annual rainfall and covering one-third of the land surface, are an important habitat for bacteria. Research on desert populations of microbes is still at an early stage.

The resilience of bacteria is legendary. They can fall into a kind of prolonged sleep and survive for millions of years. And, of course, they live on and with us—as guardians of our skin, as helpers in our gut, as assistants for our immune system, as germs, and as household aids. Without them, there would be no sourdough, no sauerkraut, no yogurt or kimchi.

The diversity of bacteria is also astonishing. They can be shaped like spheres, rods, commas, filaments, stars, or cubes. There are bacteria with flagella (tiny whiplike appendages for movement) and others without. Some are solitary and others form colonies. Bacteria can communicate with one another, join forces, and exchange genes. They can "steal" genetic information from other organisms, and they can even integrate fossilized DNA into their own genetic material.

For a long time, most people knew bacteria only as disease-causing germs. The fact that they were capable of triggering devastating plagues was firmly anchored in the collective memory, but thanks to vaccinations and antibiotics, the role of bacteria in deadly illnesses seems to have been almost forgotten. It is well known that bacteria inhabit our gut and are essential soil organisms, but most people would be amazed to learn that they also inhabit the paper and glue of this very book.

To date, more than 14,000 species of bacteria are known, with some 1,400 being pathogenic—capable of causing disease. Nobody knows exactly how many species exist, but estimates from recent research on seawater and the seafloor indicate that there could be as many as a trillion different bacterial species. That's roughly five times the number of stars in our galaxy.

The selection presented in this book cannot even begin to illustrate the incredible diversity of bacteria. Even the current record holders—the largest and smallest bacteria; the most resilient to acids, alkalies, heat, or cold; and so on—could well be dethroned tomorrow. In the coming years and decades, we can be absolutely sure that other species with exceptional capacities will be discovered or developed by synthetic biology.

But first, what is a bacterium? And how do we learn about these organisms?

Discovery of an invisible world

Despite their ubiquity, and although bacteria were used in ancient times for producing and preserving foods, they were only discovered 350 years ago. The simple reason is that, with very few exceptions, they are invisible to the naked eye.

The first person to see bacteria was probably the Dutch cloth merchant and scientist Antonie van Leeuwenhoek, born in 1632. It's unclear how he became interested in microscopes, but his interest was so great that he abandoned his business and became a natural scientist. Through his microscopes, he observed phenomena being discussed at that time, such as blood cells flowing in veins and the various developmental stages of organisms like frogs and mosquitoes. During his long life, he made several hundred microscopes and even more lenses. With magnification up to 270 times and perfect resolution, his microscopes were far better than any other instrument available at that time and enabled observation of the tiniest objects with unparalleled clarity and definition.

In August 1683, Van Leeuwenhoek fell ill. His tongue had a furry coating, making it difficult to taste foods and prompting him to investigate his sense of taste with various herbal and spiced infusions. Three weeks later and fit again, he found a forgotten glass with a pepper infusion that had become cloudy. He inspected the liquid under his microscope and discovered what he called *animalcules*— tiny animals—moving vigorously. These were probably small bacteria. He then turned his attention to his mouth and observed a sample of dental plaque. Again he discovered animalcules. He made drawings and prepared slides, together with detailed descriptions of his studies, and mailed them to the Royal Society in London, England, on September 17, 1683.

We know all this because in the early 1980s, British biologist and author Brian J. Ford wanted to find out more about Van Leeuwenhoek and visited the Royal Society's archives. There, he found nine packets, all carefully wrapped in four layers of brown paper. To his surprise, he found not only letters and drawings but also thirty-nine microscopes and hundreds of slides that Van Leeuwenhoek had made during his life and sent to London.

Studying the well-preserved slides using modern techniques revealed that Van Leeuwenhoek really had observed bacteria typical of dental plaque.

However, Van Leeuwenhoek's discovery didn't cause much of a stir in the seventeenth century. The microscope did indeed make the tiny organisms visible, but with the technology of the time, they couldn't be properly examined. Magnification was too low to reveal more than the organisms' shapes, and the ambient temperature caused them to quiver to such an extent that they repeatedly sprang out of focus. Today, this phenomenon is known as Brownian motion.

And what were these strange rods, spheres, and hooks? Were they animals or plants, or maybe parts of plants? Were they just different forms of one and the same organism? Could they transform from one into another? And, most importantly, could such tiny organisms have any significance at all for humans?

These were not unreasonable questions at a time when many facets of life, how it originated, and how it developed were still largely not understood.

Until well into the nineteenth century, the consensus among medical professionals was that transmittable diseases were caused by noxious vapors known as miasmas. This line of thought wasn't particularly outlandish, as the air in towns and cities was contaminated by decaying garbage, rotting offal from slaughterhouses, stinking cesspools, and the smoke from hearth fires and workshops. There were no sewage systems, nor organized garbage disposal. According to the beliefs of the time, people in the countryside were infected by fumes from manure and putrid swamps.

Even royal palaces were not spared the stench. The social historian Alain Corbin, in his book *The Foul and the Fragrant: Odor and the French Social Imagination*, published in 1986, cites contemporary sources writing of Versailles:

> The unpleasant odors in the park, the gardens, even the château, make one's gorge rise. The communicating passages, courtyards, buildings in the wings, corridors, are full of urine and feces; a pork butcher actually sticks and roasts his pigs at the bottom of the ministers' wing every morning; the avenue Saint-Cloud is covered with stagnant water and dead cats.

At that time, most naturalists believed that certain organisms were continually being spontaneously generated from the influences of heat, air, water, and cosmic forces on nonliving matter—mice from old rags and grains of wheat, fleas and lice from the sweat of humans and dogs, and so on.

Long disregarded and then feared

Almost two hundred years after Van Leeuwenhoek, and following much bitter controversy, the research of French scientist Louis Pasteur and others established that food spoilage, as well as alcohol fermentation and other processes, could be attributed to microscopic organisms killed by heat or chemicals.

Pasteur's first ingenious experiment involved pouring a clear boiling broth into a sterilized glass bottle with a swan's neck. Once the bottle was full, he heated the tapered neck over a flame to melting point and sealed it. The broth in the bottle remained clear as long as the bottle was sealed. As soon as the neck was broken, allowing access to germs from the air, the broth went off within a few days. One of his sealed bottles remains and is exhibited at the London Science Museum. The broth that it contains is clear and unspoiled to this day.

Pasteur disproved claims that the broth's deterioration was the result of chemical effects from atmospheric gases by sealing the bottles with cotton wool or by heating the neck and extending it while adding a curve. These methods allowed air to enter but not microbes from the outside, since all particles settled either on the cotton wool or before the kink in the bottle's neck.

Pasteur was able to convincingly demonstrate not only that microbes didn't spontaneously generate and were the cause of decaying foods, but that various microbes had different physiological capabilities. Some microbes could produce alcohol, others acetic or lactic acids.

Thus, people began, bit by bit, to understand the significance and spread of Van Leeuwenhoek's "tiny animals" and eventually recognized that they were neither animals nor plants but their own distinct realm of life—what we now call a domain. Systematic study of bacteria and other microbes had begun.

At that time in the nineteenth century, the term *bacteria*—ancient Greek for "staff" or "cane"—and eventually other terms like *bacillus*, *coccus*, and *spirillum* ("rod-shaped," "round," "spiral") were introduced to differentiate various shapes. In everyday usage today, however, the word *bacteria* is used for all shapes of bacteria. The word is also generally used to include archaea, which were first discovered in the 1970s, even though, like eukaryotes (plants and animals) and bacteria, the archaea are in their own domain.

During the "golden age of bacteriology" toward the end of the nineteenth century, people finally learned to fear bacteria as a cause of epidemics. In 1897, the illustrated weekly magazine *Die Gartenlaube* (The garden arbor), with a readership of one million, included a much-noted article written by "Dr. St." and entitled "Mikrokokken und Bacterien" (Micrococci and bacteria), which said:

In every breath of air we take, with every sip of water, and in many foods that we ingest as nourishment for the maintenance of life, thousands upon thousands of such creatures migrate into our organism.

The sudden realization of the ever-present threat of bacteria occupied writers like Alexander Moszkowski. In his poem "Überall Bakterien" (Bacteria everywhere), published in the magazine *Fliegende Blätter* in 1887, and translated here, he wrote:

Na, come off it! Bacteria?
Times used to be much cheerier,
When eatin' was a pleasure

With drinkin' yarns to treasure.
But since they found bacillus
And other shapes that kill us,
The human race is a-sufferin',
Our health as good as nuffin'.

Exhibitions depicted the image of bacteria as a many-headed Hydra; popular scientific books explained the struggle against bacteria—humanity was at war with epidemics and their instigators. German physician and microbiologist Robert Koch paved the way for combating epidemics and infectious diseases by devising principles of hygiene and was soon hailed as a hero. The portrait of this great opponent of bacteria adorned commemorative coins, cups, beer mugs, and pipe bowls. No other branch of science gained such great medical and social significance so rapidly.

Simultaneously, scientists were discovering the role of bacteria in many everyday processes—vinegar fermentation, the curdling of milk, food spoilage, the formation of humus, and symbiosis with certain plants.

How are bacteria studied?
At the advent of bacteriology as a new scientific discipline, researchers had limited resources. Their most important instrument remained the microscope, which now offered considerably greater magnification and resolution. At first, research consisted of observing bacteria through microscopes to identify, describe, and classify their various shapes. To aid this work, researchers made drawings and took photos.

The real breakthrough came when science learned how to culture bacteria. Initially, scientists cultured them on slices of potato, coagulated protein, or pieces of meat, or in broth. Growth, however, was difficult to observe, and by no means were all bacteria capable of growing on the foods on offer. Eventually, Koch's lab hit on the idea of thickening liquid foods like puddings with gelatin and letting them dry in thin layers inside test tubes and later in large rectangular bowls.

For the first time, bacteria were growing on a reasonably transparent surface, but the shape of test tubes hampered observation, and the rectangular bowls were hard to handle. The greatest frustration, however, was that gelatin becomes liquid at 98.6°F (37°C), and this was precisely the ideal temperature for culturing many pathogens.

Two further innovations from Koch's lab, both still in use today, finally enabled the rapid development of modern microbiology: agar—the gelling agent made from red algae—and the petri dish. The first innovation came from an idea that occurred to Fanny Angelina Hesse, a scientific illustrator and lab assistant, who knew of agar from her hometown of New York. She knew agar was used as a thickener and realized that the algal extract was more stable at higher temperatures than gelatin. Another of Robert Koch's assistants, Julius Petri, had the idea of using two round glass dishes of different sizes. He poured a thin layer of thick, hot nutrient medium into the smaller one and placed a dish with a slightly larger diameter over top to act as a lid.

Now, dishes with culture media could be prepared in advance and stored without spoiling. Bacteria could be cultured in situ to form colonies—clusters of cells, visible to the naked eye, that emerged from individual bacteria. These are technically referred to as pure cultures. The shape and color of the bacterial colonies made identification possible, a method still used in hospitals today to identify pathogens. Microscopes and petri dishes thus helped to establish the involvement of bacteria in diseases, in the fermentation of foods, and in many other processes.

Finally, scientists discovered that bacteria could be stained, allowing them to distinguish bacteria from dead organic matter. The Danish physician Hans Christian Gram experimented with a variety of dyes, and in 1884 stumbled across one that colored bacteria in animal tissues. His method didn't work on all bacteria, but it was nevertheless an important way to classify them. Even today, we distinguish gram-positive from gram-negative bacteria. The difference in coloration has to do with the composition of the cell wall, with gram-positive bacteria taking up the crystal violet stain used in the test and appearing purple under a light microscope. With this information, bacteria can be classified into two broad categories according to their cell walls.

Until well into the twentieth century, it was impossible to see inside bacteria with the available instruments, because the resolution achievable with a light microscope was limited by using visible wavelengths of light. In the 1920s, electrons were discovered to have wave characteristics, and like light, they could be broken. This paved the way for the development of electron optics. The wavelengths of these matter waves are around four orders of magnitude shorter than those of visible light. Thus, the era of electron microscopy began. Modern electron microscopes allow us to see inside bacterial cells, and even to see individual molecules.

In the second half of the twentieth century, physicists developed microscopy processes that helped overcome the traditional limitations of light microscopy. Today, using molecular biological and biochemical techniques, researchers can investigate modes of action, control mechanisms, and the interaction of individual structures and molecules within bacterial cells. Films taken in labs show how bacteria ingest food, exchange genes, invade foreign cells, and much more. A video taken by a British-Japanese research team, for instance, shows in real time how a bacterial enzyme attacked and severed the threadlike genetic material of a virus.

However, many problems remain: one is the inaccessibility of many bacteria. For one thing, they often live in extreme habitats. It is immensely difficult to take samples from the stratosphere, Antarctica, the floor of the deep sea, hot or acidic springs, and rocks miles below the Earth's surface and then to keep bacteria from these environments alive for lab research.

Another problem is cultivability—a prerequisite of studying bacteria in a lab is being able to culture those bacteria. In principle, it should be possible to culture all types of bacteria, but in practice only a tiny fraction, maybe 1 percent, can be cultured currently. The remainder live or grow under conditions that cannot be reproduced at a reasonable cost.

A third problem is the overwhelming number of bacteria. There are many thousands more bacterial cells on Earth than grains of sand, and the estimates of how many species exist vary widely. Every year, a few thousand new bacterial species are discovered, but probably not even 1 percent of those on our planet have yet been discovered and described. Estimates vary between one hundred billion and as many as a trillion species of bacteria.

Today, bacteriology relies heavily on molecular biology. Bacteria are identified by sequencing their genetic material—that is, deciphering their genetic code rather than observing them through a microscope or on petri dishes. Instead of identifying new bacteria one by one, researchers isolate the entirety of the genetic material from a sample. Subsequently, the composition of this "metagenome," which contains the genetic material of thousands of different bacteria, is analyzed. With the aid of genomic libraries and algorithms, researchers can then identify both known and new species. Computer analysis provides information about which known species the newly identified bacteria are closely—or not so closely—related to, which enzymatic tools they have at their disposal, and how and where they probably live. From these analyses, it is even possible to deduce whether a newly discovered bacterial species is motile or can form spores. With this method, scientists have recently been able to identify hundreds of new bacteria solely in the human gut—which was thought to have already been well researched.

In the search for new medicinal substances, researchers analyze environmental samples to extract complete genetic materials from bacteria. The individual sequences of metagenomes are then sorted by size and each one integrated into known bacteria. Finally, tests are run to find out whether these genetically modified bacteria produce a new substance with a desired effect on other cells. The search for new enzymes to produce or break down commercially relevant substances works in a similar way. This work is done with machines that use extremely small samples in the tiniest of vials.

Modern methods are extremely efficient but as a result, bacteria as organisms have slipped from the limelight. Today, very few bacteriologists observe them under a microscope, and there is little funding for time-consuming cultivation trials. Bacteria are no longer classified and described by their appearance, shape, and habits but instead by their genetic code. Scientists publishing papers on novel bacteria often don't even mention the size or appearance of their cells but rather the size and structure of their genome—that is, the genes within the objects of their studies. However, it becomes more difficult to assign functions to unknown genes when there's no way to compare with well-documented cultured organisms.

New methods like high-throughput cultivation should remedy this difficulty. High-throughput cultivation allows cost-effective processing of many samples in parallel and on a very small scale, so that organisms can then be cultured on a larger scale and analyzed in the classical way.

Nevertheless, even with the best lab equipment, it can still take months to successfully grow a new species. Despite the huge advances in molecular biology, a prerequisite of bacterial research remains cultivating and isolating bacteria along with submitting new species to an official, publicly accessible library of bacterial strains.

Laboratory culturing is not the philosopher's stone. As the word *isolation* indicates, extracting bacteria from their habitats means removing them from natural interactions with their surroundings and other bacterial communities. We lose the opportunity to observe microbial communities and their communication with one another. Therefore, in recent years, we've seen the emergence of the field of microbial ecology, where researchers study whole symbiotic communities. Their most important tools are shovels, bores, and pressurized coring tubes with which they can remove intact layers of sediment, soil, and water. Samples are then examined not only biologically and molecularly, but also using techniques from chemistry, physics, and geology to determine sample composition and the processes at work. In this way, researchers can better understand the complex interactions among bacteria, other organisms, and geological factors.

Does a rose by any other name smell as sweet?

The use of Latin in naming bacteria follows the traditional conventions of early taxonomy. According to the designations introduced by Carl Linnaeus in the eighteenth century, the first part of the name always identifies the organism's genus and begins with a capital letter. This is followed by an epithet or tag in lowercase, which, in combination with the genus, characterizes the species. For instance, with *Bacillus thuringiensis*, the genus is *Bacillus* ("stick" or "rod") and the specific name is *thuringiensis*, referring to Thuringia, the place where this bacterium was first isolated. There are at least two hundred different species in the genus *Bacillus*. A close relative of *B. thuringiensis* is *B. anthracis*, the cause of anthrax; its specific name is Greek for "coal," a reference to the black skin lesions typical of the disease.

The word *Candidatus* (in italics) precedes the genus name until a bacterium is maintained in a culture collection. Among biologists, the (unwritten) right to name a new genus or species traditionally belongs to those who first describe it. There are no limits to the imagination. Some bacteria are named in honor of fellow scientists, others after the place where they were discovered, and yet others after particularly striking features. *Legionella shakespearei* was found in Stratford-upon-Avon, the birthplace of William Shakespeare. A close relative, *Legionella gratiana*, received its epithet from the Roman emperor Gratian, who was fond of bathing in the hot springs where the bacterium was first discovered. The genus *Basfia* is named after the chemical corporation BASF, because the first isolated strain of the genus was made in one of its labs. The slime-forming bacterium *Deefgea rivuli* gets its name from the German Research Foundation, DFG (de-ef-ge), and *Propionibacterium acnes zappae* pays homage to the musician Frank Zappa. The Italian microbiologist who first described the bacterium, Andrea Campisano, is a passionate Zappa fan. He reported his astonishment at discovering that the bacterium he had just isolated from a grapevine belonged to the family of acne bacteria, and at that very moment, he was listening to the *Sheik Yerbouti* album with Zappa singing about sand-blasted zits.

The origins of life

There is clear evidence that organisms were well established on Earth 3.5 billion years ago, a billion years after the planet's formation. In 2019, Australian scientists used electron microscopes to study 3.5-billion-year-old sedimentary rocks of the Dresser Formation in Western Australia, and they identified distinct microbial residues—just like those produced by bacteria today. This eureka moment literally robbed the lead researcher, Raphael Baumgartner, of a night's sleep.

We assume that at that time, bacteria—or rather, their ancestors—had the Earth to themselves. Reports from speleologists, cave explorers, going into undisturbed karst caves millions of years old suggest how the world of these bacterial ancestors might have looked. The walls of these caves are coated with moonmilk, or inch-thick layers of pale bacteria and their whiteish residues. Snottites (from the English word *snot*)—long, sticky structures with the consistency of nasal mucus that are made of bacteria and threadlike strands of slime—hang like stalactites from the cave ceilings. The floor is crumbly and soft because bacteria have transformed it into an oatmeal-like mass.

The microbial residues found in Australia suggest that the organisms creating those structures had already had a long history. Thus, microfossil discoveries and indirect evidence from even older geological formations indicate that life might have originated 4.3 billion years ago, only a hundred million years after the first oceans were formed.

According to many biologists, the oceans—or, more precisely, the areas around hot springs on the ocean's floor—are the likely spots for the origin of life on the planet. These hot springs are formed when subterranean magma comes into contact with ocean water. The water heats up, is enriched by minerals, and shoots at high pressure and temperatures of up to 570°F (300°C) through fissures in the seafloor. Such hydrothermal vents, known as "black smokers," can be found today in all oceans. The first organic molecules could have come from these chemistry labs vigorously mixing acids, minerals, gases, and water, and could then have formed larger and more complex structures. The details of how this actually happened remain a mystery.

The evolution of life from then on seemed to be straightforward: Over millions of years, the first single-celled organisms emerged—the prokaryotes, to which bacteria belong—and later, the eukaryotes emerged. The key difference between these two groups is in their genetic material. For prokaryotes, it is readily accessible inside the cell, while for eukaryotes, the material is within a nucleus, separated from the rest of the cell by a membrane. Unicellular organisms like amoebas and paramecia, as well as all fungi, plants, and animals, including human beings, belong to the eukaryotes.

Inside the cells of all organisms with a nucleus—that is, all eukaryotes—are small organs called organelles that resemble bacteria in size and shape. Mitochondria are the organelles responsible for generating energy. Green plants also have chloroplasts, the organelles that conduct photosynthesis. Both mitochondria and chloroplasts contain genetic material with similar structures to bacteria, so scientists think these organelles are linked to bacterial remains that were in a symbiotic partnership with the precursor to eukaryotes. This symbiosis would have been akin to the symbiosis between algae and fungi in lichens, or between rhizobia (nitrogen-fixing bacteria) and certain plants. Initially, the symbiotic bacterial partner would have been inside cells, and in the course of evolution, the bacteria lost most of their genes to the genome in their partner's nucleus. You can see different stages of this kind of symbiosis between higher organisms and bacteria to this day as bacteria accumulate on cells, then migrate into and live inside cells, and eventually start to lose their own genetic material within the nuclei of those cells.

In the 1970s, the firmly established idea of these two domains of life changed. The first to raise doubts was American microbiologist Thomas D. Brock as he studied living communities in the sulfurous hot springs of Yellowstone Park. Brock and undergraduate student Hudson Freeze were surprised to discover pink mats that seemed to be made of bacteria in the nearly boiling hot zones of Mushroom Spring. They isolated a microbe, which they described as a new type of bacteria, and named it *Thermus aquaticus* (living in heat and water).

Many researchers initially doubted that an organism was able to live at a temperature of 185°F (85°C), since the upper limit was thought at that time to be 163.4°F (73°C), but the discovery eventually made history in two ways. First, the organism's enzymes, stable in extreme heat, launched the field of biotechnology, because they facilitated replication of genetic material in the lab through a technique called polymerase chain reaction (PCR). PCR was invented in the 1980s by Kary Mullis, who was awarded a Nobel Prize for the invention. Second, and maybe even more importantly, the discovery of *T. aquaticus* started the hunt for other exotic bacteria. In quick succession, more and more species were discovered living in seemingly extreme habitats or with unusual metabolic characteristics.

Such bacteria fascinated microbiologists like Otto Kandler and Karl Stetter in Germany and Carl Woese in the United States. Stetter and Kandler analyzed and compared the cell walls of bacteria because they were interested in primordial bacteria, which were thought to have had only primitive cell walls, if any at all. Woese, on the other hand, compared the genomes of organisms, or more precisely, highly complex clusters of certain ribonucleic acids (RNA) that form organelles called ribosomes. Present in the cells of all living organisms, ribosomes function a bit like a 3D printer to produce proteins. A messenger RNA (mRNA) is formed in a cell's nucleus as a working copy and takes information to the ribosomes to make the needed protein. Ribosomes were first discovered in 1960, and by about ten years later, scientists knew that the genetic sequence of the ribosomal RNA is highly conserved, meaning that there is comparatively little difference in the RNA of animals, plants, and bacteria, unlike the genome in the cell nucleus.

Woese speculated that he could make conclusions about species' common ancestors and how far back they could be traced by examining variations in ribosomal RNA. He was convinced that this information would make it possible to identify a history or tree of life.

Woese, Kandler, and Stetter got to know one another and exchanged bacteria and data. The fact that many of the newly discovered exotic bacteria had cell walls different to any other known bacteria supported Woese in his hunch that these organisms, despite their appearance and behavior, were not bacteria at all. His team continually thought up new ways of determining the genetic sequence of the ribosomes in these "exotics" and comparing them with "normal" bacteria and other organisms.

When Carl Woese and his then research associate George E. Fox (later an astrobiologist and a professor at the University of Houston consulting for NASA) published their findings at the beginning of November 1977, the *New York Times* ran an article a few days later with the headline "Scientists Discover a Form of Life That Predates Higher Organisms."

Woese, a solitary man who didn't like traveling, avoided conferences, and was not a good speaker, is now considered the most important biologist of the twentieth century, although his name is almost unknown outside professional circles.

We have since learned that these life-forms—originally known as archaebacteria but now known as archaea—form a third domain of cellular life, the other two being bacteria and eukaryotes. Like bacteria, archaea have no nucleus and reproduce by cell division when they reach a critical size. Among the features they share with eukaryotes are specific mechanisms to regulate and replicate genes that are completely different from systems in bacteria. Researchers now believe that we are more closely related to archaea than we are to bacteria.

Bacteria probably diverged first from one common universal ancestor, which biologists have termed LUCA—last universal common/cellular ancestor. It's unclear and still controversial whether RNA originated first and DNA later, such that LUCA belonged to the RNA world, and whether there ever was such a thing as a common ancestor. In the early days of life on Earth, the exchange of genetic material could have been so frequent that it was more of a primordial collective.

There is agreement, however, about when the domains Archaea and Eukaryota parted ways, because domain

Bacteria had already split off. In time, archaea, specializing in extreme habitats (heat, acids, alkalies, low nutrients, etc.), lost certain characteristics and gained new ones. Eukaryotes became more complex and developed into multicellular organisms with strict division of tasks for individual cells and tissues, allowing organs to evolve. Researchers believe archaea initially specialized in extreme habitats as a way to escape viruses. Presumably, the early phases of life were more heavily influenced by stray naked genes and early viruses, neither of which were well adapted to withstand extreme conditions.

We still cannot explain a remarkable phenomenon—archaea colonize plants and animals, even living in our guts, but despite this, not a single species of archaea has yet been found to cause disease in humans, other animals, or plants.

One thing is certain: roughly 2.7 billion years ago, there were already clearly distinguishable bacteria and archaea, and 400 million years later, bacteria evolved a form called cyanobacteria that use sunlight for photosynthesis. Cyanobacteria multiplied so quickly that the atmosphere became enriched with oxygen, causing the Great Oxidation Event, in which oxygen gas, being toxic to most organisms at the time, triggered a mass extinction of species. This exceptionally fast-moving event also changed the color of the Earth's surface by oxidizing many rock formations from black to rust red, and caused a rapid decline in the greenhouse gas methane, resulting in an ice age. This led to the further extinction of species, including many oxygen producers, causing oxygen levels to decrease and allowing the world's temperatures to rise again.

At around 1.9 billion years ago, so after another 400 million years had passed, the first eukaryotes evolved. In this evolutionary lineage, the kingdoms of protists, fungi, plants, and animals formed.

What can bacteria do?

Bacteria and archaea are by no means primitive organisms that came to an evolutionary standstill. They are highly complex and have the capacity to quickly adapt to adverse conditions. Their adaptability enables them to colonize extreme habitats (excessive heat, acids, alkalies, salinity, heavy metals, lack of air, etc.) that eukaryotes cannot. They can be observed today settling rapidly, even in clean rooms, and also in disinfectants or places that are lethal for eukaryotes such as chemically contaminated or radioactive environments.

This is where they benefit from the modular construction of their genomes. Individual functional areas are, as it were, centralized and jointly regulated. The "housekeeping" genes for basic metabolic functions and the genes responsible for cell shape hardly differ between species, but there is a wide variety of genes for special or additional functions. Most genes are in the form of a DNA double helix, about as long as a one-cent coin is wide. The helix is tightly twisted into a closed ring, the chromosome. Most bacteria have only one chromosome, and a very few have two. Plasmids—mainly circular and considerably smaller DNA double helixes that float freely in the cell—provide bacteria with more flexibility. They usually contain genetic information for specific metabolic pathways of importance for survival in stressful situations. As an example, they may contain the genetic information to break down substances toxic to a bacterium such as antibiotics or enable them to metabolize unusual molecules. Bacteria can actively exchange plasmids or absorb them from their environment—for example, after the decay of other bacteria. Plasmids are the reason that antibiotic resistance can spread rapidly among bacteria and can also be passed on to other bacterial species.

Although they are single-celled organisms, bacteria have numerous specialized structures. Some bacteria contain vacuoles for the storage of gases used for metabolism or for buoyancy, giving the organism the ability to move up and down in aquatic environments. Others have vacuoles or inclusion bodies—"containers"—to break down or store certain molecules. Then there are structures such as magnetosomes that help certain bacteria orient in the magnetic field.

On the exterior of some bacteria are whiplike structures called flagella, which facilitate targeted movement in liquids. They are the only known biological structures that form a truly rotating joint, functioning like a propeller. The rotation frequency is around 40–50 Hertz. A chemically powered mini-motor in the cell membrane drives each flagellum, while an array of sensors and molecular switching mechanisms control its rotational movement. Some bacteria can even change the direction of a flagellum's rotation, rather like a ship's propeller.

Sensors reacting to light, magnetism, or chemical substances help orient the bacteria to the presence of nutrients or surfaces. Pili—small, sticky, hairlike appendages made of protein—help attach bacteria to nutrients or surfaces, including to plant and animal cells, and they are often crucial for the pathogenicity of bacteria. Special conjugation pili, also called sex pili or F pili (F for "fertility"), allow bacteria to exchange genetic information. The donor bacterium attaches its F pilus to the recipient bacterium and draws the two cells together. Once the distance between the two cells is sufficiently small, the cell membranes temporarily fuse and form a channel through which genes can be transferred. This is not true sexuality, but it does allow DNA transfer.

The exchange of genetic material is not a requirement, however, for bacteria to reproduce. Under favorable conditions, bacteria can grow and multiply very rapidly by a process called binary fission. Once a bacterial cell reaches a certain size, its genome doubles. Then the two genomic loops separate and migrate to opposite ends of the cell, the cell's midsection constricts, and finally the cell divides into two. The whole process can take place in only about twenty minutes, so under the right conditions, a bacterial population can double in size within that time—a fact that should be kept in mind when leaving perishable foods lying around at room temperature.

However, not all bacteria reproduce by binary fission. Some grow to more than twice their original size and then undergo multiple divisions to produce multiple daughter cells. Some reproduce by budding—in which a bud forms on the mother cell and grows until it is large enough to separate and become its own cell, and still others form daughter cells internally until the mother cell gives "birth" to the new bacteria.

In the mid-twentieth century, scientists found that the exchange of genetic material can also occur between two unrelated bacterial species. This discovery has been significant both for medical sciences and for ideas about how life has developed. It also offers an explanation for the rapid adaptability of bacteria and archaea. Exchange between species doesn't even require pili—sometimes viruses participate, and sometimes bacteria absorb genetic material released by other bacteria, a bit like absorbing food from their environment. Bacteria can release genes to their surroundings when they're destroyed, perhaps from viral attacks or by other sudden changes to their environment.

The phenomenon of gene transfer involving viruses was first observed by Victor J. Freeman, an American doctor, in the 1950s. He was studying *Corynebacterium diphtheriae*, the pathogenic bacteria causing diphtheria, to understand why some strains that were unable to trigger symptoms of the disease suddenly became virulent, causing life-threatening infections. Freeman used viruses to characterize the *C. diphtheriae* strain—a standard practice in bacteriology. The particular viruses he chose were bacteriophages ("bacteria eaters"), which attack and infect bacteria. Most bacteriophages are highly selective and can infect only certain strains of bacteria. Therefore, when bacteria are cultured in petri dishes and infected with a selective bacteriophage, the breaking up of the bacterial colony can be seen with the naked eye. To his surprise, Freeman detected that some previously nonvirulent *C. diphtheriae* bacteria became virulent after he had infected them with bacteriophages. It turned out that the virus had transmitted the gene needed by the bacteria to produce the toxins that cause diphtheria.

In 1959, Japanese scientists discovered that different species of bacteria could exchange genetic material with each other, enabling them to become resistant to certain antibiotics. The rapid transfer of resistance factors between various bacteria has, in the meantime, become a serious global problem in the use of antibiotics to fight bacterial infections. Increasingly, pathogens are becoming resistant to one or more antibiotics, and there are already pathogenic bacteria that have become resistant to all available antibiotics.

The transfer of genetic material not only takes place between various species of bacteria, but has also been observed between different organisms by a process known as horizontal gene transfer. Bacterial genes have been discovered in fungi, higher plants, worms, insects, and higher animals; plant genes have been found in fungi, bacteria, and animals; and even human genes have been transferred to single-celled organisms.

Geneticists believe that the human genome has a considerable number of genes from other organisms, and about forty of them have a bacterial origin.

For bacteria living in anaerobic environments, with no access to oxygen, an average of about 16 percent of the genome comes from horizontal gene transfer. Taking into account only those genes responsible for metabolism, the proportion increases to almost one-third. In contrast, only about 8 percent of genes are foreign in aerobic bacteria, living in environments with access to oxygen. The reason for this difference is still unclear. Here, too, most of the foreign genes are involved in adaptation and in exploiting new or multiple energy sources.

The adaptability of bacteria, as a rule, has less to do with mutation than with incorporating already existing genes or genetic elements.

Horizontal gene transfer also plays a significant role in the development of archaea. For instance, in *Methanosarcina* archaea, the genes that enable them to convert acetates into methane and carbon dioxide originate from cellulose-digesting *Clostridium* bacteria.

These are not just academic findings; horizontal gene transfer has had devastating effects on many organisms all over the world. A sudden mass extinction of species—the Permian–Triassic extinction event—occurred 252 million years ago, and within a window of a few tens of thousands of years, almost three-quarters of terrestrial animal species, including many insects, died out, as did countless land plants. In the oceans, 95 percent of all invertebrate species died. All in all, about 90 percent of species living at that time went extinct, including 99 percent of all vertebrates.

While the oxygen content of the atmosphere prior to the beginning of this catastrophe was 30 percent (today it is 20.9 percent), afterward it declined to between 10 and 15 percent, mainly because of the global extinction of plants.

Originally, scientists thought this extinction event could be traced to a series of volcanic eruptions in what is now Siberia releasing so much carbon dioxide that dramatic changes were triggered in the atmosphere and climate. These volcanic eruptions are among the largest known volcanic events in Earth's history; they carried such huge amounts of copper, nickel, and palladium from the interior of the planet to the surface that these metals are still being mined in Siberia today.

In 2012, however, a research trip to China by American geophysicist Daniel Rothman changed scientific thinking about the mass extinction. Rothman analyzed sediment cores from this geological era and found that the greenhouse gases had increased far too quickly to be related to geological processes.

Only the influence of living organisms was an alternative worth considering. Rothman's team analyzed the genome of *Methanosarcina* archaea, which, even today, are responsible for the bulk of methane emissions from biological sources. To their surprise, the scientists discovered that these archaea acquired the gene for methane production at precisely the beginning of the extinction event. The concurrent release of nickel by the volcanoes added a necessary ingredient—to function, the methane-producing enzyme system of *Methanosarcina* needs nickel.

The explosive growth of archaea—which, thanks to gene transfer and geological events, were able to exploit the ecological niche of breaking down organic substances into gases in deep, anaerobic ocean waters—caused a massive release of methane and carbon dioxide. The consequences were acidification of the oceans and, within a few tens of thousands of years, a greenhouse effect that dramatically changed the climate.

In other respects, too, bacteria have turned out to be transformative. The recent discovery of bacterial immune systems is of tremendous scientific and technological importance. Bacteria and archaea, like higher organisms, can be infected by viruses. Bacteria can defend themselves, as they have developed an immune memory that can be passed on to their daughter cells. If bacteria successfully survive a viral infection, gene segments of the virus are integrated in their genetic material. The bacteria store the viral gene segments by flanking them with special sequences—clusters of short DNA segments that can be read forward or backward like palindromes (such as the words *madam* and *racecar*). Numerous copies of these DNA palindromes are spaced in a regular arrangement, hence their abbreviated name as the acronym CRISPR (clustered regularly interspaced short palindromic repeats). The spacers in between the palindromic repeats—the viral gene segments—are, therefore, the central elements of the bacterial immune memory and allow the bacteria to fight new infections by these viruses.

This part of the mechanism serves as the immune memory. The CRISPR region with the fingerprint of the virus can be reread when necessary and translated into RNA. This RNA, because of the palindromic sequences, then folds into a looplike structure with the viral sequence attached to it as a kind of "wanted" poster.

Now the Cas protein (*Cas* from "CRISPR-associated") comes into play as a further element of the bacterial immune system. When a new viral infection occurs, the Cas protein forms a complex with the RNA sequences transcribed from the immune memory, binding to the loop structure while leaving the section with the "wanted" poster of the virus free. This free segment attaches to the matching sequence in the viral genome and once the binding occurs, the adjacent Cas protein clips and thereby inactivates the genome of the viral invader. At the location of the cut, a random sequence is inserted, further contributing to the deactivation of the viral genome. In this process, the RNA segment acts like a sniffer dog to lead the Cas protein to the appropriate place on the viral genome, and the Cas acts like a pair of scissors, cutting a gene at precisely the sequence defined by the "sniffer dog." Thus, the common description of CRISPR/Cas as "gene scissors" is quite fitting. As the "memory" elements are found inside the bacteria's genome, they are passed on to daughter cells with every cell division.

Since the discovery of further variations of the CRISPR/Cas system, more enzymes have been identified. All are able to shut down individual genes in a targeted manner, and they are part of the natural mechanisms used by bacteria and archaea to inactivate genes from viral intruders.

This offers exciting new ways to understand the biology of bacteria and interactions between bacteria and viruses. Especially transformational is the possibility of exploiting the immune system in archaea and bacteria for biotechnology. Using CRISPR/Cas, it is possible to make changes at a precise location in a gene. At the site of the cut, a cell usually inserts a random sequence, making the change indistinguishable from a natural mutation. This way, targeted mutations can be induced in many organisms, as the enzymes work well not only in archaea and bacteria, but also in plant and animal cells. These mutations can activate or deactivate certain enzymes and thereby change metabolic pathways.

Scientists can also exploit the mechanism to insert new gene sequences at the site of a cut, introducing the genetic material at specific places in any organism. In the past, a genome could only be changed at random, but now, using genome editing, many new applications are possible in medical sciences, agriculture, and chemistry. An array of identification segments and Cas proteins can be ordered at low cost from commercial suppliers, so biologists all over the world have an inexpensive, quick, and easy-to-use tool at their disposal with which to rapidly and precisely transform organisms' genetic material. At the moment, almost on a weekly basis, new insights, methods, and applications for this genetic surgery are being published.

Why are bacteria important?

Bacteria didn't only have a profound effect on the planet's atmosphere; they also provided the prerequisites for plant and animal life and thus for the development of the human species. The role bacteria play in metabolic and energy cycles is also enormous, if only because their numbers are virtually inconceivable.

According to estimates, there are four to six nonillion bacteria on the planet, a sum that by far exceeds the estimated number of stars in the universe (seventy sextillion). When written out, one nonillion has thirty zeros: 1,000, 000,000,000,000,000,000,000,000,000. Scientists abbreviate that by writing 1×10^{30}. A sextillion is 1×10^{21}.

Bacteria exploit dissolved salts, minerals, dead organic materials, and many other chemical compounds, and in doing so, they produce nutrients for plants and animals. They can convert nitrogen gas from the atmosphere into ammonia so that organisms have soluble nitrogenous compounds available to produce proteins. The ability to transform atmospheric nitrogen—called nitrogen fixation—is a unique feature of bacteria and archaea. Conversely, some bacteria can convert nitrates back into nitrogen gas. Therefore, they have a huge role in the nitrogen cycle. The same applies to the sulfur cycle. Certain bacteria convert sulfide created by the decomposition of organic matter to sulfates needed by plants and animals. Bacteria are involved in the carbon cycle, too.

Cyanobacteria, also known as blue-green algae, profoundly changed the Earth's early atmosphere when they developed the ability to capture carbon dioxide from the atmosphere between 2 and 4 billion years ago. In the process, oxygen was released and over the course of several billion years, the growth of cyanobacteria and of the green plants evolving from them changed the carbon dioxide-rich atmosphere of early Earth to decrease the amounts of carbon dioxide and increase the amounts of oxygen. Cyanobacteria still account for about 20 to 30 percent of global photosynthesis and are an important element of the carbon cycle. Other bacteria release carbon dioxide into the oceans and the atmosphere; for example, deep-sea bacteria dissolving carbon-containing rocks or soil bacteria breaking down organic matter.

Bacteria also play a part in the release of other carbon-containing greenhouse gases such as methane. For these reasons alone, it is important to study how and where bacteria live and to understand which bacteria release gases in specific ecosystems and under what circumstances.

For plants and animals, bacteria have a significant role in supplying nitrogen and in digesting cellulose-rich food in the stomachs of ruminants. Added to this is their role in the development of diseases in plants and animals.

Humans, too, have an intimate relationship with bacteria. First, they live on and in us—bacteria colonize our skin and mucous membranes, teeth, stomach, intestines, lungs, and very possibly our brain, too. In the intestines, they assist with food digestion, but their role on and in other organs is still unclear. On the other hand, bacteria cause infectious diseases and devastating epidemics such as plague, cholera, and tuberculosis.

Bacteria are also important for the production of vinegar, the preservation of vegetables and other foods, and the production of dairy products like sour cream, yogurt, and cheese. Recently, they have become vital for producing a host of other substances. In the last three decades, the chemical industry has undergone a transformation that has largely gone unnoticed by the public. Medicines, vitamins, dyes, food supplements, and many basic chemicals are now produced by bacteria. With bacteria involved in the process, manufacturers can make these substances faster, using less energy and without producing toxic intermediates. Instead of operating complex manufacturing plants with each synthesis cycle needing dozens of steps at varying temperatures, pressures, and so on—all of which affect yield—today's manufacturing plants consist of steel tanks similar to the drums used to brew beer. Inside these tanks live bacteria producing the desired product from raw materials. Often, it's sufficient to use bacterial enzymes rather than bacteria themselves (the word *enzyme* is derived from the ancient Greek for "in yeast"). Almost all powdered and liquid detergents today contain bacterial enzymes, which ensure that starch, fat, and protein residues are broken down, that grass and wine stains disappear, and that fibers become smooth—all at 104°F (40°C) and, in the near future, even in cold water. The enzymes, in turn, are broken down by other bacteria in the wastewater. In Germany alone, detergent enzymes save 1.5 million tons (1.4 million metric tons) of carbon dioxide annually.

In the textile industry, bacterial enzymes are responsible for environmentally friendly bleaching and the stone-washed effect, and in leather processing, they are used to clean animal skins. Producers of natural cosmetics, foods, and animal feed use bacteria-derived products to avoid the use of solvents or to replace synthetic substances.

Finding bacteria with new attributes is not difficult. Bacteriologists imagine, as it were, being bacteria and take samples from garbage landfills, hot or acidic sources, contaminated ground, saltwater lakes, and sometimes even from the parking lots of their own workplaces, just to see whether the organisms they find could, under the right conditions, produce or break down a certain substance. In this way, bacteria have been found that digest dioxins and plastics, and others that extract carbon dioxide from coal-burning power plant emissions.

Other technologies are being used to optimize enzymes for industrial purposes. In a process copied from nature, bacterial enzymes are adapted step by step to specific parameters, such as temperature, presence of solvents, or pH values.

Last, but not least, the CRISPR/Cas technology already described offers numerous applications in medicine, agriculture, and chemistry.

Bacteria in distant worlds

The world is teeming with bacteria, so are they present elsewhere, beyond Earth?

This question arises because of several facts. First, bacteria can live in extreme temperatures, both hot and cold, and in acidic and alkaline environments. They can survive high pressure, vacuums, and even harsh radiation. They survive not only space flight but also unprotected exposure in space.

Second, their spores can survive for millions of years without any nutrients and still retain the ability to germinate. Theoretically, bacteria, or rather their spores, could have survived in boulders catapulted into space by some planetary catastrophe. Using the example of the asteroid that crashed into our planet 66 million years ago and caused the extinction of dinosaurs, scientists have calculated that this collision hurled large amounts of rock from Earth into space, with the debris circling the sun in irregular orbits. Many of these pieces of debris have likely landed on Mars, just like Martian rocks catapulted into space from similar events have landed on Earth. Some pieces of debris from Earth might even have reached the moons of Saturn and Jupiter. Calculations and simulations suggest that many of these objects would have contained bacteria or bacterial spores. Therefore, it is entirely plausible that life could be transported from one celestial body to another by cosmic events.

Third, bacteria can exist without sunlight. Life on the Earth's surface is entirely dependent on sunlight—plants grow thanks to photosynthesis and feed countless animals; the animals, in turn, are prey for other animals and their remains feed all sorts of organisms in soil, lakes, rivers, and oceans. In the dark abysses of the deep sea, organisms live on the constant supply of dead material drifting down like snowflakes.

There are, however, bacteria in and around deep-sea hydrothermal vents, deep within rocks, and in other habitats (e.g., in lakes beneath million-year-old glaciers, and in certain cave networks) that obtain their energy solely by oxidizing inorganic compounds. There is a possibility that life first began under these conditions, with the sun-dependent communities emerging much later. Over the course of evolution, entire communities that are neither directly nor indirectly dependent on sunlight have developed around deep-sea vents.

Thus, even celestial bodies far away from the sun could support life. Some planets and moons have temperature ranges from below freezing to as high as 265°F (130°C), which some microbes on Earth are known to tolerate. Examples of such places include Mars; the upper layers of Venus's atmosphere; Jupiter's moons Io, Europa, and Ganymede; and Saturn's moons Enceladus and Titan. So, it is quite conceivable that life exists in those places, too, or that bacteria that somehow landed there could have gained a foothold.

For the time being, however, we know of life only on Earth, and this life provides plenty of puzzles: How did it all start, and how did it develop to the complexity seen today? Which coincidences and catastrophes led to the development of plants and animals, the extinction of dinosaurs, the evolution of elephants from shrew-like mammals, and the conditions for primates to emerge and eventually become humans? How many small and large events were there, how many dead ends, mass extinctions, and convergent evolutions?

Higher plants and animals, including humans, have a long and complex pathway behind them. The heritage of countless ancestors is coded in our genes. In contrast, microbes like bacteria and archaea are among those organisms that have, essentially seamlessly, survived all the untold twists of evolution. True, they have continuously adapted and over millions of years repeatedly specialized, but in essence, their basic blueprints have remained the same. This allows great insights into the diversity of life that can result from variations of so few elements.

At the same time, studying these organisms is providing surprises time and again.

Many insights have transformed our understanding of life and evolution. First, there aren't two domains but three: Bacteria, Archaea, and Eukaryota. Second, it is now clear that plants, animals, and humans are more closely related to the newly discovered Archaea than they are to Bacteria. Third, we know today that microbes are much more complex than biologists even in the twentieth century thought possible. The discovery of a bacterial immune system, as already mentioned, has contributed greatly to this notion. The fourth insight is the discovery of horizontal gene transfer. This is particularly exciting, as it does away with the Darwinian concept of a tree of life with clear branches that can be used to trace who is descended from whom. In reality, the branches tangle with one another and sometimes rejoin or form cross-connections, even over long distances—that is, between organisms, including humans, that are evolutionarily distant or are even from different domains.

The fifth and final insight is that although humans and bacteria inhabit the same planet, they live in completely different worlds. For us, the world is bright, warm, wide open, and airy, while for most bacteria, it is dark, cold, low in oxygen (hypoxic), and small. Their living spaces are extremely limited, even though they can be transported over long distances from continent to continent, in air, by water, or in another organism. Because of their small size, bacteria find water more viscous than syrup would be for us. When their flagella stop beating, their movement stops after a distance equivalent to an atom's thickness. With a limited energy supply, growth and reproduction often come to a standstill. Thus, bacteria have to be not only great survivors but also masters of adaptation to cope with crises and acquire new sources of food and energy. They have transformative powers and the potential to change entire planets.

In the next section of the book, I provide insights into the diversity of life among bacteria and archaea with portraits of fifty of these fascinating organisms.

Record Holders

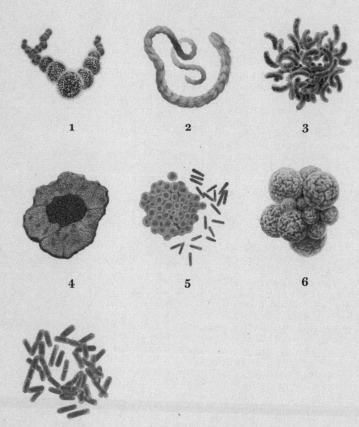

1. *Thiomargarita namibiensis*
2. *Eubostrichus dianeae* epibacterium
3. *Pelagibacter ubique*
4. *Nasuia deltocephalinicola*
5. *Minicystis rosea*
6. JCVI-syn3.0
7. *Lysinibacillus sphaericus*

Thiomargarita namibiensis
Namibian sulfur pearl

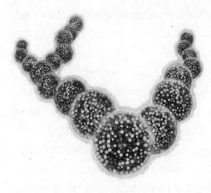

Diameter: 100 to 300 micrometers (µm), sometimes larger
Growth form: Often chains of spherical cells; under certain conditions, cells can have an elongated shape

THERE IS A GERMAN SAYING that trees don't grow to the heavens; actually, trees can reach a maximum height of only about 425 feet (130 m). It's a physical factor that governs this limit: trees cannot actively transport water. Only when water evaporates from the leaves does the pressure drop sufficiently to allow water to flow upward from the roots. This mechanism can function to a height of 425 feet, but no higher.

For animals, gravity and energy balances set the size limits. With bacteria, the governing factor is the rate of diffusion—the speed that nutrients and gases can move within a cell.

It was all the more surprising, then, to discover *Thiomargarita namibiensis*, because at first glance, this bacterium seems to contradict the basic laws of physics. With a diameter sometimes reaching 750 µm, it is roughly the same size as the full stop at the end of this sentence. It's currently the largest known bacterium and one of the few visible to the naked eye. If the average bacterium were the size of a caterpillar, then *Thiomargarita* would be the size of a blue whale. But only about 2 percent of its volume is used for a particular function or for reproduction; the rest of the cell consists of storage space. So, in effect, the giant only appears to be a giant.

The Namibian sulfur pearl was discovered in 1997 by German microbiologist Heide N. Schulz and her team while on board the Russian research ship *Petr Kottsov*, where they were inspecting sediment samples from Walvis Bay off the coast of Namibia. The researchers were actually on the trail of other sulfide-oxidizing bacteria, but the sulfur pearl kept appearing in their samples, and its size was so extraordinary that initially Schulz and her colleagues couldn't believe that they had stumbled across a new bacterium. Since then, *T. namibiensis*—or rather, its close relatives—have been found in other maritime regions too.

T. namibiensis forms long, branched chains of four to twenty, sometimes even fifty, cells held together by a mucous sheath. Sulfur inclusions refracting light lend the cells a pearly luster.

The cells of *T. namibiensis* are large because they have to be well stocked. Where they live off the Namibian coast, the seabed is particularly rich in sulfides, which they oxidize by breaking down nitrates in a process accomplished without the help of oxygen. With this ability, *T. namibiensis* has tapped into a source of energy that other bacteria in anaerobic habitats are unable to exploit.

But to get the nitrates for this process, *T. namibiensis* cells need contact with salt water containing nitrates. The sulfur pearls live in the sulfur-rich seabed, often buried by thick mud, and they have to wait until methane eruptions or storms stir up the mud and bring them into nitrate-rich water. Then, they quickly take in large quantities of nitrate and store it in vacuoles, which resemble inflated balloons and can account for up to 98 percent of the total volume of their cells. Once the bacteria become buried again, they can live for a number of years using these nitrate stocks, until a storm whips up the sea and churns the sediments to once again bring the sulfur pearls into contact with water. Nitrate concentrations in the inner cells of *T. namibiensis* are ten thousand times greater than in salt water.

While in contact with seawater, *T. namibiensis* also takes advantage of the available oxygen to stock up on phosphate, which the cells use to generate energy once they're isolated from water. Since the bacteria occur at such high densities off Namibia, their phosphate turnover is high enough for the bacteria to form phosphate minerals called apatites on the seabed. Therefore, by extracting soluble phosphate from the seawater, these bacteria play an important role in the phosphorus cycle.

It was recently discovered that *T. namibiensis* can also come in different shapes. In the Gulf of Mexico, researchers found these bacteria not in the usual pearl necklace form but in clusters of two to sixteen cells looking rather like a cauliflower. And off the Pacific coast of Costa Rica, *T. namibiensis* was found in threadlike cell structures attached to hard surfaces; these cells regularly budded daughter cells. Despite these different cell shapes and structures, most of the interior of these giant cells consists of stored-up reserves.

Eubostrichus dianeae
epibacterium
Bacterium living on Diane's bushy-haired roundworm

Length: Up to 120 μm
Width: 0.4 μm
Primary source of energy:
Oxidation of sulfur

THE EXISTENCE OF the *Eubostrichus dianeae* epibacterium also seems to contradict physicochemical principles. With a length of up to 120 μm, it is the longest known bacterium. It seems fair that it also has one of the longest common names—the bacterium living on Diane's bushy-haired roundworm.

These bacteria live on the outer surface of the *E. dianeae* roundworm and in such great numbers that they give the worm the appearance of a Wookiee, the shaggy fantasy creature in George Lucas's popular *Star Wars* series. The *E. dianeae* epibacterium, which has yet to be formally named, is a threadlike gamma proteobacterium related to *Thiomargarita namibiensis*, described in the previous profile.

Each *E. dianeae* worm carries as many as 60,000 of these bacteria, which together account for up to 44 percent of the creature's total volume. The bacteria are each attached from one of their ends to the skin of the roundworm, and they are arranged in such a tight, regular order that they lend a furlike look to the worm, which always seems to be well groomed. Individual bacteria can grow to about one-tenth of a millimeter and can be seen by the naked eye. Thus, they are the longest known bacteria that are still able to divide. Like *T. namibiensis*, the *E. dianeae* epibacteria can oxidize sulfur. Sulfur can be found in the form of grainy deposits between the cytoplasmic membrane and the outer membranes of the bacteria.

How can molecules move through the long distance within each bacterium? Diffusion of nutrients and the movement of molecules within the cell would take too long. However, it's possible that long-distance molecular movement isn't actually needed for metabolism in these bacteria, because the interior of each cell seems to be divided into sections. Having separate internal compartments may help the bacterium overcome the problem of its length. This hypothesis gets support from the fact that each of these long cells has up to sixteen copies of its genome.

According to the discoverer, the roundworm is named after a certain Diane Curtis—her identity remains a mystery. Sometimes the worm's name is wrongly spelled *E. dianaea* or *E. dianae*.

The *E. dianeae* roundworm was first discovered in the 1970s in the sea off southern Florida, where biologists and geologists were analyzing sulfurous sediments. The worm can also be found off the Caribbean coast of

Central America. It feeds on organic waste covering the seabed. The bacterial fur coating is a form of symbiosis and is common to all known *Eubostrichus* epibacteria species that oxidize sulfur and store elemental sulfur. Some *Eubostrichus* worms are completely hairy, and others only partially so. Zoologists believe that the roundworms get nourishment from the bacteria, and that the bacteria benefit as the worm squirms through the sediment and brings them into contact with food and energy sources.

Unlike for other known *Eubostrichus* roundworms, the epibacteria on *E. dianeae* attach only one of their ends to their host's surface. Despite the length of each bacterium, when they multiply, they divide exactly at their center. What happens to the half not attached to the worm remains a puzzle, as does the reason for their extreme length, which makes them a record breaker among bacteria.

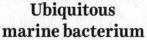

Pelagibacter ubique
Ubiquitous
marine bacterium

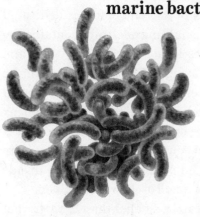

Shape: Mostly crescent-shaped rods
Length: 0.37 to 0.89 µm
Width: 0.12 to 0.20 µm

IF THE *ESCHERICHIA COLI* bacteria in our gut were the size of a rabbit, *Pelagibacter ubique* would only be as big as a mouse. This ubiquitous marine bacterium is not only the smallest independently living bacterium, but also the most efficient and the most successful living organism on the planet. Every liter of salt water has millions of them. There are an estimated 10^{27} to 10^{28} *Pelagibacter* cells in the world—one hundred thousand to one million times more than the number of stars in our universe. But that isn't their only record. Seawaters are extremely low in nutrients, so to survive, microbes have to actively transport the necessary nutrient molecules across their membranes to the cell interior. This transport costs energy and, in the end, there has to be a surplus. *P. ubique* lives at the boundaries of what is possible. It has everything it needs for food intake, growth, and reproduction, and not a bit more or less.

P. ubique, of all organisms, is also a leader in the highly efficient use of space. It is almost impossible to imagine using less space to maintain metabolism and achieve reproduction. Two-thirds of its cell volume is dedicated to metabolism, and the remaining one-third contains the genetic material. Sensors locate carbon, nitrogen, and iron compounds, as well as light. The cell also has all the necessary transport systems and all the enzymes to produce the twenty amino acids essential for living organisms.

Smaller organisms are only possible if metabolism is totally or partially waived. Viruses, for example, are more or less just tightly enclosed packets of genes that invade other living cells to seize control of their metabolism.

Bacteria and archaea can be smaller than *P. ubique* but only when they can do without housekeeping genes by having a reliable source of nutrition available. One example is *Mycoplasma genitalium*, a pathogenic bacterium that can infect mucous membranes in the urinary and genital tracts. It measures a mere 300 by 600 nm (nanometers), but it is unable to live independently. In 2015, researchers claimed that they had found even smaller bacteria living in groundwater, but none of them have been cultured and most experts doubt their existence.

The genome of the *P. ubique* "survival machine" is also a miracle of efficiency. It has only 1.3 million base pairs and contains almost 1,400 genes. No other known independently living organism manages with fewer genes. Nothing is redundant in the genome; it has only the absolutely necessary components. Even the genetic code seems to be optimized for maximum energy use. *Pelagibacter*'s genetic code, as in all organisms, uses the four bases A, C, G, and T, but A and T are found in far greater numbers than in other bacteria. It's all about efficiency: C and G contain more nitrogen, a rare commodity in seawater, and are thus more difficult to produce. It's as though *Pelagibacter* forms its code like a text written from carefully selected words and letters using as little printer ink as possible.

P. ubique is a curiosity among its close relatives, the Rickettsiales, nearly all of which live in the cells of other organisms—many as pathogens, such as *Rickettsia prowazekii*, the cause of lice-borne typhus. Biologists study *P. ubique* not only because of its astonishingly efficient use of energy and the organization of its genome, but also because of its enormous ecological significance. All *Pelagibacter* bacteria taken together exceed the total weight of all the fish living in the sea. They make up a quarter of the bacterial biomass in the oceans, and in the warm summer months, they can be as much as a half. Since they live on soluble carbon compounds, mainly from decomposing organisms, *P. ubique* bacteria play an important role in the carbon cycle.

Their abundance also makes them attractive to enemies; several viruses have been identified that can attack and kill the bacterium.

P. ubique was only discovered in 2002. Until then, scientists had known only about its ribosomal RNA, found in water samples taken from the Sargasso Sea in 1990. The bacterium was one of the first to be detected using identification techniques that were new at the time, but attempts to culture it failed. Researchers finally succeeded by using highly diluted culture media and adding a dye that was able to attach specifically to ribosomal *Pelagibacter* RNA.

Nasuia deltocephalinicola
Socho Nasu's leafhopper bacterium

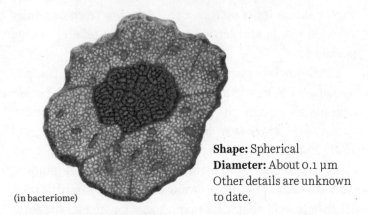

(in bacteriome)

Shape: Spherical
Diameter: About 0.1 µm
Other details are unknown
to date.

FOR INDEPENDENTLY LIVING BACTERIA, having 1,400 genes is probably the lower limit. To get by with fewer genes is only possible if certain capabilities aren't needed. *Nasuia deltocephalinicola* demonstrates just how far this can go. This microbe, which lives in leafhoppers—tiny, sap-sucking insects—holds the record for the smallest genome ever found in a bacterium. It has 112,000 base pairs and only 137 genes coding for proteins.

Leafhoppers feed off the sap produced by plants. Because plant sap is predominantly rich in carbohydrates but lacks proteins and other nitrogenous building blocks, the leafhoppers have to carefully manage the sparse supplies of nitrogen so they can make the proteins needed for growth. To do this, they maintain two different strains of bacteria in a specialized organ called a bacteriome. Inside the bacteriome are specific cells called bacteriocytes that envelop the bacteria. Unlike gut bacteria, they are trapped in a cellular cage that they can only leave when the leafhopper reproduces.

From the meager sources of nitrogen in the leafhopper's gut, *N. deltocephalinicola* produces two amino acids; another eighteen are produced by *Candidatus Karelsulcia muelleri*, a second bacterial species living in the bacteriome. This bacterium, like the Russian matryoshka dolls, contains yet another bacterium, called *Arsenophonus*.

When a female leafhopper becomes sexually mature, all three bacteria penetrate the egg-producing cells. As soon as the eggs are fertilized, have divided, and have grown into larvae, the bacteria settle in a cell that will later become the new bacteriome. This mechanism works so reliably that it has long been used by viruses that migrate from generation to generation under the protection of these endosymbionts.

Fossil finds show that the symbiosis between leafhoppers and their three nutrient-providing bacteria has existed for at least 260 million years. This is the conclusion reached by molecular geneticists studying different species of leafhoppers and their respective endosymbionts; they were able to determine when the last common ancestor of the bacteria and leafhoppers must have lived. Today, the originally independent bacteria lack essential components of their metabolism. Similar endosymbioses have been found in all sap-sucking insects investigated to date. Sometimes fungi are involved. Some insects house bacteria that can exploit nitrogen from the air. And sometimes, in the course of evolution, genes of the bacteria became completely incorporated into the insect's genome.

Researchers have compared the genomes of various bacterial endosymbionts and identified eighty-two genes common to all of them. They believe that these are probably the "essential housekeeping genes" that are vital to the existence of bacteria. In addition to these genes for maintenance of basic cellular functions, there are genes for certain services that the bacterium provides to its host. The absolute minimum threshold for such bacteria could be ninety-three genes, requiring roughly 73,000 base pairs.

The integration of bacteria and their genes will probably continue over the next millions of years. At some point, the leafhoppers may have an organ that produces nitrogen with the help of bacterial genes. Biologists would then likely not find any foreign organisms in this organ.

Working with these seemingly exotic bacteria is not only of interest for basic research purposes. It could also have enormous practical consequences, because without these symbionts, the sap-sucking insects—including many agricultural pests (cicadas, locusts, etc.)—cannot exist. So, if plants could be cultivated that are harmful to these bacteria, then the pests would die or be made infertile once they had eaten that plant. This would provide a way to target a particular insect pest—and only when the insect eats that particular crop.

Nasuia deltocephalinicola got its genus name in honor of the Japanese entomologist Socho Nasu. The epithet, or specific name, refers to the occurrence of the bacterium in the Deltocephalinae subfamily of leafhoppers.

Minicystis rosea
Pinkish-red blister bacterium

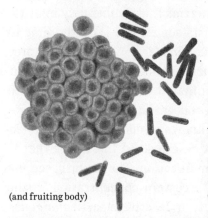

(and fruiting body)

Shape: Long, cylindrical rod
Length: 3 to 8 μm
Width: 1.2 to 1.3 μm
Fruiting body: Oval
Spores: Round
Special characteristics:
Oxygen-dependent (aerobic)

MINICYSTIS ROSEA LIES at the other end of the genome size scale. In 2014, this bacterium replaced the long-standing record holder *Sorangium cellulosum* as the bacterium with the largest known genome. With sixteen million base pairs, it surpasses *S. cellulosum* by three million. Just for the sake of comparison, the genome of an average bacterium consists of a few hundred thousand to a few million base pairs. In the genome of *M. rosea*, there are probably 14,000 genes—ten times more than in *P. ubique*, considerably more than in the previous record holder, and about two-thirds of the number of human genes.

The bacterium was discovered during a systematic search for new myxobacteria in a soil sample from the Philippines that had been tucked away for years in the specimen collection of the Center for Biodocumentation, a German research institute. Myxobacteria are interesting for several reasons. For one thing, they produce a whole range of substances with biological or pharmacological effects. For another, they show signs of having a multicellular way of life, including an ability to coordinate with each other. They form swarms that deliberately hunt for other microbes—biologists sometimes jokingly refer to these groups as wolf packs. Once they find nutritious prey, they release digestive enzymes that break down food outside the cells. The coordinated attack enables them to reach a sufficiently high concentration of these digestive juices. They use what is known as quorum sensing, a biological mechanism first described in the 1990s that is based on chemical communication. The bacteria secrete substances that, above a certain concentration in the environment, can activate or deactivate particular genes. Thus, the number of bacteria secreting these substances determines the behavior of the bacteria.

When the living conditions are unfavorable, *Minicystis* bacteria form round fruiting bodies that mature into spores. These spores show only minimal signs of metabolic activity, and they can survive, protected from dehydration and ultraviolet light, until food conditions improve.

Because it is small and round, the bacterium was named *Minicystis*—little bubbles. The name *rosea* reflects the cells' pink-to-red coloring.

M. rosea's diet has not yet been fully clarified. It can break down other bacteria with its digestive juices, and it can produce steroids, an unusual feature among bacteria. These abilities make *M. rosea*, like many other myxobacteria, interesting for biomedical and industrial use. Many substances produced by myxobacteria are suitable as precursors for the production of more complex chemicals or are being tested for use as drugs in the fight against cancer and infectious diseases.

Whether the size of its genetic material is connected in some way to its social abilities, meaning its high degree of complexity, remains unclear. There must be a reason, because the larger the genome, the more complex and energy-consuming it is to maintain and replicate during cell division.

With bacteria and archaea, there is usually a reasonably linear relationship between genome size and the number of genes. The larger the genome—the number of base pairs making up the organism's DNA—the more genes that particular microbe will have.

This relationship does not exist in higher organisms. Humans have some 20,000 genes, while the considerably less complex water flea has 31,000 genes, making it the current record holder among animals. The situation is similar with higher plants. After decades of genome research, it is clear that, for eukaryotes, biological complexity has nothing to do with the size of the genome or the number of genes. This is a puzzle that shows we still have not fully understood how the complex interplay between genomes and other factors really works. For this reason alone, *M. rosea* is a worthwhile object of research.

JCVI-syn3.0

Synthetic bacterium from the
J. Craig Venter Institute, version 3.0

Shape: Spherical
Diameter: About 0.7 µm

BACTERIA ARE WELL SUITED for investigating the structure and function of genomes and comparing them with the more complex structures in eukaryotes. By studying bacteria, we can also learn about the minimum requirement for an organism to live independently. That challenge, along with previous knowledge of the modular construction of bacterial genomes, gave scientists the idea to construct a minimal organism, now known as JCVI-syn3.0. This bacterium, whose name sounds more like a software product, is an artificially produced organism with only 473 genes, fewer than any other free-living bacterium. It is an intermediate step on the way to *Mycoplasma laboratorium*, a bacterium that doesn't yet exist. The idea is for this future bacterium to contain only those genes that are essential for functioning, forming a basic structure to which specific abilities can later be added—rather like constructing a commercial vehicle with a chassis, motor, and gears as a platform onto which a cab, tipping mechanism, refrigeration unit, and other extras can be placed. *M. laboratorium* is the equivalent for synthetic biology, intended to provide a platform in which the necessary infrastructure exists for growth and reproduction. Later, this basic genome can be used to add functions to create organisms for particular tasks, such as extracting carbon dioxide from the atmosphere; producing hydrogen, drugs, and other commodities; decomposing toxic waste; exploiting metals from waste; and so on.

The team at the J. Craig Venter Institute (JCVI) in La Jolla, California, took *Mycoplasma* bacteria, known for their extremely small genome, as a starting point. Most *Mycoplasma* are parasites of humans, animals, and plants. In people, they cause a form of pneumonia and urinary tract infections, among other problems. They are so small that they aren't visible under a light microscope, and since they live as parasites, they have lost their cell wall in the course of evolution.

The researchers began with the pathogen *M. genitalium*, and with innumerable steps, they systematically destroyed each of its 482 genes, one at a time, and looked at the effects. In a publication, the researchers described 382 of them as being essential. But *M. genitalium* grows very slowly, so the researchers turned to the closely related bacterium *Mycoplasma mycoides*. They chemically produced its 1,017 genes using the digitally available gene sequence through synthesis automatons and transplanted them into cells of the bacterium *Mycoplasma capricolum*, whose genome had previously been removed.

The synthetically created *M. mycoides* bacteria, called JCVI-syn1.0 or Synthia, showed the expected characteristics and were able to proliferate.

In a next step, based on JCVI-syn1.0, another attempt was made to create a functioning organism with a minimal genome: JCVI-syn3.0. This first truly synthetic bacterium has less than half of the original 1,017 genes and grows about three times slower than syn1.0, but it is viable, forms colonies, and behaves like natural bacteria. The analysis of its morphology, metabolism, and gene regulation is ongoing. Of the 473 genes it contains, one-third are not yet known to have any function. This surprising finding demonstrates how little is known so far about the functions that are vital for the life of organisms.

Another complete genome of an organism that was produced entirely synthetically is *Caulobacter ethensis-2.0* (the specific name refers to ETH Zurich, a Swiss institute of technology). However, an associated functional organism doesn't yet exist. The synthetic genome is based on *Caulobacter crescentus*, a freshwater bacterium that scientists study because its cell division results in two different daughter cells. Unlike the *M. mycoides* genome produced at the Venter Institute, the *C. ethensis-2.0* genome was not a one-to-one copy but was simplified on the computer without altering the actual genetic information. To do this, the researchers removed so-called redundancies in the genome. The genetic code contains certain three-part codes (called triplets or codons), which are similar to synonyms in language. For example, the triplets UUA, UUG, CUU, CUC, CUA, and CUG all code for leucine, an essential amino acid. Any one of these codes can be used to advise the cell's ribosome machinery to add leucine to the protein under construction, just like the word *also* in a text could also be expressed with the words *too, further,* and *as well.*

In the end, roughly one-sixth of all 785,000 base pairs were modified. Initial tests of the individual genes revealed that more than 80 percent of the newly described essential genes were functional.

These findings give researchers hope that creating organisms with synthetic genomes may be easier than was originally thought.

In 2019, scientists at the University of Cambridge reported the creation of an *Escherichia coli* bacterium with a synthetic genome. They, too, removed all redundancies from the genome on a computer—18,214 in all. The simplified genome was then synthesized step by step and gradually inserted into *E. coli* bacteria until the natural genome was completely replaced. The new organisms, which the researchers named Syn61, multiply just like their natural models, but the speed of cell division is somewhat slower, and the organisms with the artificially produced and simplified genomes are a bit larger.

The paring down of the genome certainly has practical applications. The bacterium *E. coli* (page 212) is used in industry as a production organism, and viral attacks can lead to expensive production losses. Thanks to the recoding, though, viruses can't exploit the cells as easily for their own reproduction.

Scientists plan further procedures to use the omitted codons for giving new instructions to the cell, such as integrating artificial amino acids into proteins. This might create proteins with entirely new properties that could be used as pharmaceuticals or industrial enzymes. Other research teams have already succeeded in integrating amino acids not present in nature into *E. coli*.

These plans have attracted public attention, because producing such organisms is seen by some as being at odds with religious beliefs. But, looking at it objectively, these bacteria are merely the continued development of production organisms that has been done in labs and industrial plants for decades to produce aromas and flavors, vitamins, amino acids, and drugs.

In all of these production organisms, the genome has been considerably modified through numerous interventions. It is optimized to produce the required chemicals and manipulated in such a way that the bacteria cannot survive outside labs or cultivation tanks. Some processes use genes from the human genome, so that human hormones like insulin can be produced. People with diabetes all over the world inject insulin without fear of their body having an immune reaction, as was the case with the previously used porcine insulin. Are these production organisms therefore human-bacteria chimeras? Do they have human qualities?

The same applies to crops, livestock, and domestic animals that we humans have changed beyond recognition. We bred broccoli, brussels sprouts, and kale from wild cabbage, and dog breeds as diverse as the Saint Bernard, Pug, and Chihuahua from the wolf. Researchers took decades to identify the original corn plant, and our wheat is the result of crossing at least three kinds of grain. The only difference is that those changes took so long that they seem natural to us. But can our sense of time be a moral standard?

Lysinibacillus sphaericus

Rounded bacterium containing lysine

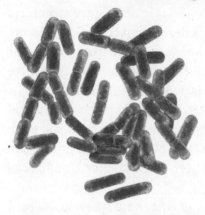

Shape: Rod
Length: 1.9 to 2.3 μm
Width: 0.6 to 0.7 μm

MULTICELLULAR ORGANISMS HAVE a wide range of life expectancies—from the one-day wonder of the mayfly (whose short existence, however, is preceded by larval stages lasting several months) to Galápagos tortoises that can live for up to 177 years. The current record holder is a 507-year-old Icelandic clam, whose age can be determined by counting the growth rings on its shell. Yet, all these organisms have one thing in common: they age and die.

This doesn't happen to bacteria. They grow and divide, and the offspring grow again to divide, and so they continue, on and on. Of course, bacteria can be destroyed by sudden changes to their environment, but they don't age. Furthermore, they have developed an adaptation process to survive difficult times: they switch to a dormant state by reducing their water content and almost completely shutting down their metabolism. This helps them to survive drought, heat, chemical challenges, and even ultraviolet and cosmic radiation.

How long they actually remain viable, however, is a question that has preoccupied scientists since structures were found in three meteorites—Allan Hills (ALH) 84001, Shergotty, and Nakhla—which some scientists have interpreted to be the fossilized remains of bacteria. All three meteorites came from Mars, hurled into space by a cosmic catastrophe, most likely the impact of a large asteroid. Millions of years later, they landed on Earth. Could meteorites transport bacteria from planet to planet? This panspermia hypothesis has been discussed for a number of decades now.

Currently, *Lysinibacillus sphaericus*, previously classified as *Bacillus sphaericus*, is considered the record holder for the longest sleep. It survived in amber inside the gut of a bee that was trapped in tree resin some 25, or maybe up to 40, million years ago. When that happened, the world was populated by huge "terror birds"—ratites hunting their prey with the speed of a cheetah—as well as by an early genus of horse the size of today's fox and by the first primates that were roughly the size of today's domestic cat.

In 1995, millions of years after the bee became trapped in the amber, scientists were able to provide convincing evidence that the spores they grew on culture medium were indeed from the bee's gut and not from contamination from the local surroundings. Based on the gene sequence, they were able to prove that the bacteria resembled *L. sphaericus*, a widespread, highly versatile, and adaptable soil bacterium that still inhabits the guts of bees and helps them digest food. The genetic differences to the *L. sphaericus* of today confirm the believed age of the amber.

In 2000, another team of scientists claimed to have isolated a bacterium from salt crystals that formed 250 million years ago. They named the resurrected microbe *Bacillus permians* and, as is common practice, deposited registered specimens in a bacterial library. But other researchers reexamining the bacterium raised concerns. They found that *B. permians* was genetically almost completely identical to *Salibacillus marismortui*. In the meantime, it has also been shown that the salt dome from which the sample originated, like many others, is periodically permeated by water. So, it's highly likely that the bacterium penetrated the salt crystal in much more recent times.

L. sphaericus got its name because of the presence of lysine-containing building blocks in its cell wall. The specific name, meaning "spherical," is a reference to its spores, which are resistant to high temperatures, chemicals, and ultraviolet radiation and are well known for their longevity. The bacterium is lethal to some insects, in particular the larvae of mosquitoes. In much the same way as *B. thuringiensis* (page 167), it produces a toxin that blocks certain vital receptors in the insect's gut so that the insect dies as a result. *L. sphaericus* is therefore used to control mosquitoes in many countries. As the bacterium also binds heavy metals, scientists are testing its use for remediation of contaminated soils. *L. sphaericus* already has commercial applications in the textile industry in neutralizing residues of azo dyes from wastewater, an ability it has thanks to enzymes that can break down complex organic substances.

UPDATE: In 2020, researchers were drilling through 250 feet (75 m) of sediments at depths up to 19,000 feet (5,800 m) some 1,400 miles (2,250 km) off the northeast coast of New Zealand. There, they isolated a 100-million-year-old community of bacteria and archaea lying dormant. The researchers fed these microbes carbon and nitrogen and were able to bring them back to life. Following their 100-million-year sleep, it took only about two months for most of the microbes to increase in size and to divide. In the end, the researchers were able to revive over 99 percent of them.

Extremists

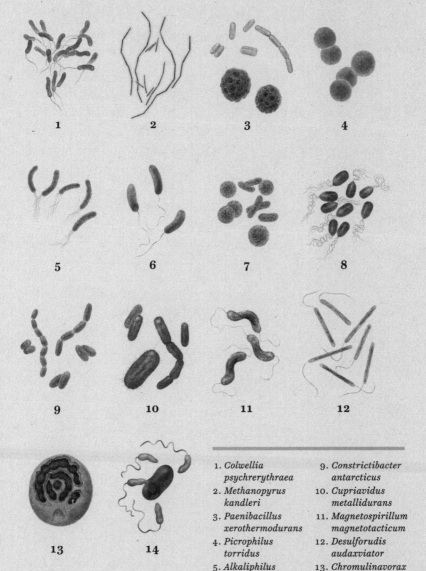

1. *Colwellia psychrerythraea*
2. *Methanopyrus kandleri*
3. *Paenibacillus xerothermodurans*
4. *Picrophilus torridus*
5. *Alkaliphilus transvaalensis*
6. *Shewanella benthica* DB21MT-2
7. *Janibacter hoylei*
8. *Halobacterium salinarum*
9. *Constrictibacter antarcticus*
10. *Cupriavidus metallidurans*
11. *Magnetospirillum magnetotacticum*
12. *Desulforudis audaxviator*
13. *Chromulinavorax destructans*
14. *Bdellovibrio bacteriovorus*

Colwellia psychrerythraea

Rita Colwell's reddish, cold-loving bacterium

Shape: Rod
Length: 2.5 to 3.5 μm
Width: 0.5 μm
Color: Slightly reddish
Locomotion: With flagellum

ACCORDING TO CURRENT KNOWLEDGE, in the early days of life on the planet, periods of great heat with average temperatures of 122°F (50°C) alternated with periods of extreme cold in which the Earth was almost completely covered in ice. Volcanic eruptions and the impact of meteorites and asteroids raised temperatures, while chemical and the first biological processes removed carbon dioxide from the atmosphere and allowed the Earth's surface to cool down.

From the perspective of most living creatures, the Earth today is a wet and, above all, cold habitat. Seawater covers more than 70 percent of the planet, two-thirds of which is cold deep-sea water with a temperature of around 35°F (2°C). Fresh water, which accounts for only 2.5 percent of all water on Earth, is not really that different in terms of temperature: 90 percent is bound up in the polar ice caps and in glaciers.

The coldest temperature measured since records began was minus 128.6°F (-89.2°C) in Antarctica; admittedly, the temperatures there rarely rise above freezing. Far more significant are places where warm periods alternate with very cold periods during the night or in winter, like, for instance, some places in Asia where temperatures reach as high as 120°F (49°C) and then sink as low as minus 58°F (-50°C). So it isn't at all surprising that some bacteria have managed to adapt to these swings in temperature.

Among bacteria that remain active at low temperatures, *Colwellia psychrerythraea* stands out. This microbe is quite happy to actively swim around at temperatures of 14°F (-10°C) and can grow and divide at temperatures as low as minus 4°F (-20°C). Researchers have even been able to detect metabolic activity at minus 320°F (-196°C), the temperature of liquid nitrogen. *Colwellia* bacteria manage to incorporate amino acids into their cellular components while in liquid nitrogen—which instantly freezes flowers into fragile glass.

Cold-adapted (cryoprotective) polymers and enzymes operating outside the cell make this possible. The bacterium covers itself with a sort of molecular netting, like a wool sweater, which prevents the formation of orderly crystal structures in the surrounding water. The cell wall of cold-adapted bacteria resembles the structure of liquid crystals and remains liquid even when it is very cold or under pressure. This phenomenon also explains why these bacteria are resistant to high pressure.

The genus *Colwellia* was first described in 1988, and the researchers suggested naming it after American microbiologist Rita Colwell. In the 1960s, Colwell discovered that cholera bacteria, naturally present in coastal and inland waters, often use tiny, algae-eating copepods as hosts. If heat or excess nutrients cause an algal bloom, these crustaceans are attracted to the bloom and bring the bacteria with them. As a result of her research, Colwell founded an initiative for safer water supplies with techniques that were as simple as using homemade filters to prevent the spread of infections via drinking water. Later she cofounded CosmosID, a company dedicated to rapid diagnosis of bacteria in environmental samples. Because she did some of her field research in Antarctica, she was also honored with an Antarctic mountain range named after her. The specific name *psychrerythraea* is from *psychros*, the Greek word for "cold," and *erythraeus*, the Latin word for "reddish," as the bacterium is resistant to cold and reddish in color.

C. psychrerythraea can also survive in anaerobic environments and use a vast number of simple and complex carbon compounds for nutrition. Since it can break down numerous nitrogen compounds and can even use sulfur as a source of energy, the bacterium is well suited for cleaning up environmental pollution in cold climates.

It could also pave the way for new vaccines. Scientists have exchanged important metabolic genes in pathogens with genes from *C. psychrerythraea*. As a result, growth of the pathogens comes to a standstill at high temperatures, and the cells die. These weakened, heat-sensitive pathogens could be used as live vaccines to safely induce an excellent immune response before they die off in warmer parts of the body. This approach has already been successful in animal experiments.

Methanopyrus kandleri
Otto Kandler's methane fire

Shape: Rod
Length: 2 to 14 µm
Width: 0.5 µm
Locomotion: With tufts of flagella at the ends of the cell
Growth form: Single or up to ten rods forming a chain; up to seventy cells in exceptional cases

NOBODY KNOWS WHETHER life started under extremely hot or moderately warm conditions. What is certain, however, is that volcanic activity, magma rising to the upper layers of the Earth's crust, radioactivity, and the heat in the Earth's hot core have provided places with consistently high temperatures: hot springs, volcanoes, the black and white smokers of hydrothermal vents, and the zone a few kilometers into the Earth's crust where temperatures reach up to 250°F (120°C).

Numerous microbes now live in these habitats, and they not only endure the heat but actually need it to survive. Among microbes, the current heat record is held by *Methanopyrus kandleri*, an archaean. The organism was discovered in 1991 following a dive on *Alvin*, a deep-sea submersible, during which scientists took core samples from a black smoker in the Gulf of California. *Alvin* is 23 feet (7 m) long, is equipped with grab arms, and has space for two scientists and a pilot. The submarine was commissioned in 1964, and although it belongs to the US Navy, it is often loaned to companies and research institutes. By 1991, *Alvin* had already been on some spectacular missions, such as the search in 1966 for a hydrogen bomb that was lost off the coast of Spain after a military airplane collision. In 1977 and 1979, *Alvin* was involved in the discovery and exploration of black smokers, which are of interest to biologists because the communities there survive completely independently of sunlight.

Smokers, chimneylike structures, form above hydrothermal vents that are generated by rising magma below the seafloor. Mineral-rich water at temperatures reaching as high as 750°F (400°C) shoots out of fissures on the seafloor. On mixing with the 37°F (4°C) ambient waters, the dissolved mineral salts precipitate out and form solid crystals of various sizes. Where the hot water exits, larger crystals form tapering chimney structures that can grow to 80–160 feet (25–50 m) tall. Finer particles swirl around in the seawater creating a smoky effect. If the waters being emitted are rich in iron salts, then the "smoke" is dark and the vents are termed black smokers, and if the waters contain dissolved sulfates, the "smoke" is lighter and they are called white smokers.

Methanopyrus was given the specific name *kandleri* in honor of the German botanist and microbiologist Otto Kandler (1920–2017). Along with Carl Woese, Kandler was instrumental in the discovery that Archaea forms a third domain of life along with Bacteria and Eukaryota.

M. kandleri is called "methane fire" because it gets its energy by producing methane from hydrogen and carbon dioxide.

It wasn't until 2018 that Japanese researchers, doing systematic tests with a variety of pressures and temperatures, discovered that *M. kandleri* could still grow and replicate up to a temperature of 251°F (122°C) when cultured under the high-pressure conditions of its natural environment. *M. kandleri* tolerates precisely one degree more than the previous record holder, the deep-sea archaean *Geogemma barossii*, also known as Strain 121. The strain number refers, of course, to 121°C (249.8°F), which is significant because it is the temperature used to sterilize medical instruments. Until these heat-tolerant organisms were discovered, it was thought that fifteen minutes at 249.8°F in an autoclave was enough to kill all life.

M. kandleri can even survive three hours of treatment at 266°F (130°C), and as soon as the temperature drops below 230°F (110°C), its growth begins again.

The heat tolerance of *M. kandleri* and other archaea is due to a number of mechanisms. Often the salinity of these archaea is increased, giving protection from heat. Thanks to special proteins, the genetic material is tightly coiled, and there are sophisticated repair systems for heat-related damage to nucleic acids and cell structures. Chemical modifications enable the protein molecules driving metabolism to be considerably more stable and robust so that they don't clot in extreme heat. Based on certain calculations, biologists currently believe that the upper limit of heat tolerance, even with protective mechanisms, is 300°F (150°C)—but this isn't certain.

Paenibacillus xerothermodurans

Dry heat-resistant "almost bacillus"

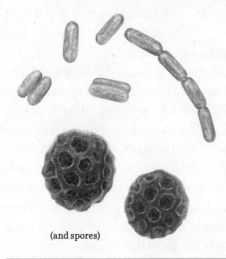

(and spores)

Shape: Rod
Length: 2.25 to 2.70 μm
Width: 0.70 to 1.04 μm
Growth form: Can live in anaerobic conditions as individuals, pairs, or chains

WHEN THE FIRST Saturn V rocket was ignited at Cape Canaveral in Florida on November 9, 1967, the five main engines—to date, still the most powerful single engines ever built—roared to life and triggered an inferno on the small island that housed Launch Complex 39. People felt the blast in their bones. Ceiling panels fell at the press center 3 miles (5 km) away, and window panes shattered 6 miles (10 km) away. Even in New York, 950 miles (1,500 km) from the launch site, earthquake monitoring stations registered the seismic tremors caused by pressure of the escaping gases. Flames longer than 1,000 feet (300 m) and at 4,000°F (2,220°C)—hot enough to melt titanium—shot out of the engine nozzles. Every second, 14.3 tons (13 metric tons) of jet fuel was burned. Although elaborate cooling systems, which released thousands of gallons (tens of thousands of liters) of water in seconds, ensured that the steel girders of the launch platform didn't melt, the ground around the platform became so hot that the vegetation charred, and bushfires started more than half a mile (a kilometer) away.

In 1973, after more than a dozen such launches—the moon-landing missions were already history—microbiologists investigated the heat-stressed ground for bacteria. Dry heat was then the gold standard for disinfecting components for space travel, and with the upcoming Viking missions that would land space probes on Mars, researchers wanted to know if any bacteria could withstand the heat treatments. Searching on the periodically scorched ground at Cape Canaveral seemed more promising than searching in deserts that are consistently hot and dry.

They did indeed find an extremely heat-resistant, spore-forming bacterium in the soil samples. It belonged to the *Paenibacillus* genus and was named *Paenibacillus xerothermodurans*, roughly translating as "dry heat-resistant almost bacillus." The paenibacilli were originally thought to be proper bacilli, until researchers discovered that they belong to a separate genus, albeit a closely related one. They were then reclassified and given the prefix *paeni-* derived from the Latin *paene* meaning "almost" or "practically."

Spores of these bacteria can withstand dry heat of 260°F (125°C) for more than ten days and moist heat of 175°F (80°C) for more than an hour. This capacity for heat resistance is due to their nine-layer spore membrane with a distinctive honeycomb pattern on its surface. Not much more is known about *P. xerothermodurans*, as its genome was only sequenced in 2018.

Paenibacillus is a diverse genus and can be found all over the world, in the tropics, in deserts, and in polar regions. The bacteria inhabit soil and water and are also found in the rhizosphere, the area directly around plant roots, where they fix nitrogen, make nutrients and trace elements available, and synthesize substances that protect plants from insects, nematodes, and pathogens causing plant diseases. Other paenibacilli are known to cause a disease called foulbrood in honeybee colonies.

The substances these bacteria produce are, therefore, of interest for medicine, agricultural crop protection, contaminated-soil remediation, and chemical production.

Picrophilus torridus

Acidophile from torrid habitats

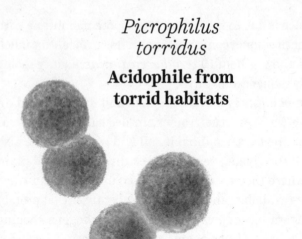

Shape: Spherical
Diameter: 1.0 to 1.5 μm

VOLCANOES, WHICH HAVE SHAPED climate and geology over long periods of the Earth's history, spew not only lava and magma but also many gases: water vapor, carbon dioxide, sulfur dioxide, hydrogen, and hydrogen sulfide, which smells like rotten eggs. When sulfur gases come into contact with water, they produce acids. Halogens such as chlorine, bromine, and fluorine, which escape together with hydrogen, also form acids. Likewise, at the Earth's hot spots and plumes, where materials from deep within the interior reach the surface and form hot springs, the water is often very acidic.

Strong acids are extremely damaging to most organisms and can readily eat holes in textiles or etch away rust from steel and iron. There are, however, bacteria and archaea that specialize in living in acids, and a number of such habitats are scattered around the world.

Near some hot springs on the Japanese island of Hokkaido, the ground is 140°F (60°C), and due to chemical compounds outgassed by magma just below the surface, it has a pH of less than 0.5—more corrosive than the acid in car batteries. This place can only be reached by people wearing protective clothing. In 1996, the archaean *Picrophilus torridus* was found in soil samples there.

P. torridus, the acid lover from a hot, dry—torrid—habitat, is a record holder among acidophilic microbes. It can grow at a pH of 0—that is, the pH value of sulfuric acid, which can destroy metals—and it can even survive negative pH values, corresponding to hydrochloric acid, which begins to smoke as soon as the flask is opened. If the acidity of the environment declines, *P. torridus* suspends growth. At the neutral pH of 7, its cells dissolve.

This incredible resistance to acids is made possible by an extremely tough cell membrane, which lets through very few positively charged hydrogen ions, called protons, and thus protects the cell interior from hyperacidity. Any protons that manage to penetrate the cell wall are transported back out by efficient proton pumps. The inside of the cell, however, is not neutral, as it is with most other acid-resistant bacteria, but has a pH of 4.6. The interior itself is so acidic that cell components and metabolism must be adapted to it. How this actually happens is not yet clear, but we do know that *P. torridus* has numerous proteins that can quickly repair damage.

Its nutrition consists of the remains of organisms that didn't survive contact with the hot, acidic environment.

P. torridus releases acid-resistant enzymes that predigest carbonic nutrients outside the cell. Numerous specialized transport proteins then deliver the individual products into the cell.

P. torridus depends on oxygen—in biology, this type of organism is called an obligate aerobe—and with 1.5 million base pairs, it has one of the smallest genomes ever discovered for an independent, nonparasitic organism. Only *Pelagibacter ubique* (page 49) has a smaller genome. Also, astonishingly, 91.7 percent of its genome codes for proteins with only the remainder being responsible for regulatory functions. On average, for the genomes of all bacteria and archaea analyzed so far, the protein-coding portion is about 80 percent.

Genetic analyses have also shown that, in the course of evolution, *P. torridus* incorporated numerous genes from bacteria and archaea in its environment. Its genetic makeup is thus optimized for survival in these extremely acidic, hot, dry, sulfate-rich soils.

Alkaliphilus transvaalensis
Alkaliphilic bacterium from Transvaal

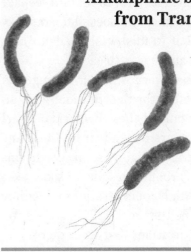

Shape: Slightly curved rods
Length: 3 to 6 µm
Width: 0.4 to 0.6 µm
Locomotion: With numerous flagella
Growth form: Singles or pairs; can also form chains of four to six cells

UNLIKE STRONGLY ACIDIC HABITATS, strongly alkaline ones are much rarer on Earth and are mostly associated with volcanic activities. Soda lakes often form in places where the volcanic subsoil remains on the surface after volcanoes or massive lava flows have cooled down. These lakes have very high amounts of sodium carbonate and are usually extremely salty.

Alkaline habitats also exist in plutonic rock, where rocks containing silicate react chemically with water and carbon dioxide. There are also examples of artificial alkaline habitats, such as sites that have been used for decades or centuries for producing leather, dyes, cement, iron, or aluminum.

Specialized microbes live in these artificial environments as well. *Alkaliphilus transvaalensis* was discovered by researchers in 2001 during a systematic search for bacteria in plutonic rocks in the pit water of a South African gold mine 2 miles (3.2 km) deep. The water came from a concrete pool that the miners had built to contain seepage from a borehole above. The bacterium doesn't tolerate oxygen and grows best at a pH of 10 (dishwashing liquid) but can still divide at pH levels above 12.5 (bleach).

A. transvaalensis is thus a record holder among alkaliphilic bacteria, which prefer growth conditions with pH levels of 9 (soapy water). Some of these bacteria only tolerate alkalies, while others are dependent on them and can no longer grow in neutral or slightly acidic conditions. They are mainly found in carbon-rich soils, in soda lakes, and in some deserts with high concentrations of sodium carbonate in the ground, and they're usually extremely tolerant of the high salt concentrations that occur there.

These bacteria all face the same problem of having to protect their interior from aggressive alkalies. Alkalies can rapidly break down cell components and nucleic acids, and an alkaline environment also affects energy production. Energy production in most cells is based on establishing an electrical voltage and an energetic proton gradient at their cell membrane by releasing positively charged hydrogen ions (protons) into the environment. This creates an energy-rich chemical gradient: the protons flow back and prompt an enzyme that provides certain energy-rich molecules for cellular metabolism. These molecules are basically the fundamental fuel of cells. In an alkaline environment, however, this doesn't work. The protons transported out of the cell react immediately with hydroxides, which are present in large amounts in the surrounding alkali, forming water molecules. This means it is no longer possible to generate energy through the return flow of protons. The details of how exactly alkaliphilic bacteria overcome this are not fully understood. Some manipulate their direct surroundings by releasing substances like acetic acid; many fortify their cell walls with teichoic acids— long-chain molecules that contribute a negative charge to the cell wall as protection against the ions of the alkaline medium. Others use specific transport molecules to exchange hydrogen ions for sodium ions, which explains the frequent presence of alkaliphilic bacteria in saline environments. Many also relocate their digestion outward by releasing enzymes into the environment.

Such external digestion can also be found in animals. Spiders, for example, inject digestive fluids into their prey, liquefying their victims from the inside to make subsequent ingestion easier. Some insect larvae, such as those of the ox warble fly, vomit digestive juices over their prey, which they wouldn't otherwise be able to reduce to manageable sizes. The same applies to sea stars. Carnivorous plants, which gain their nutrition from insects, also digest outside of their cells.

All the unusual characteristics of alkaliphilic bacteria make them interesting not only for research but also for technology and medicine. Their enzymes, being effective in strong alkalies, are well suited for detergents and for the chemical and pharmaceutical industries as a whole. The teichoic acids, which cause febrile reactions in humans by stimulating certain receptors, are useful in the production of modern vaccines, because they have a beneficial influence on the immune system's reaction to the vaccine.

Shewanella benthica
DB21MT-2
James Shewan's
deep-sea bacterium

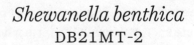

Width: 0.8 to 1.0 µm
Length: Around 2 µm
Locomotion: With a flagel-
lum at one end of the cell

TIME AND AGAIN, bacteria illustrate that conditions we
find familiar and pleasant certainly don't apply to all living creatures. This includes the atmospheric pressure
present in our environment. We define the pressure that
exists at sea level as standard pressure and quantify it as
one bar, also measured as 14.5 pounds per square inch
(PSI) and 0.987 atmospheres (atm).

This standard pressure decreases with increasing altitude and eventually is so low that the cells of normal organisms burst. Conversely, pressure increases with increasing
depth. On the floor of the Mariana Trench, 36,201 feet
(11,034 m) below sea level, the pressure is a thousand

times greater than on the Earth's surface. At this depth lives *Shewanella benthica* DB21MT-2. To date, it's the bacterium most capable of withstanding high pressure.

There is a whole range of archaea and bacteria in the deep sea at depths of many thousands of feet, but *S. benthica* DB21MT-2 not only survives this pressure, it depends on it. *S. benthica*, the originally described species, is a piezophilic (pressure-loving) bacterium that was first isolated from the sea cucumber *Psychropotes longicauda* from the Walvis Ridge, an ocean ridge in the southern Atlantic Ocean. It is named after the Scottish marine microbiologist James Mackay Shewan, and the specific name is from the Greek word *benthos* meaning "on the seabed." The additional tag of the strain DB21MT-2 indicates its origin, since the bacterium has so far been identified as being very similar to *S. benthica* from sequencing the RNA of its ribosomes. The DB21 refers to the dive (number 21), while MT-2 indicates the Mariana Trench and site where it was found. The sample was taken by *Kaiko*, a remotely operated underwater vehicle (ROV) built by the Japan Agency for Marine-Earth Science and Technology. The word *kaiko* is Japanese for "trench." Between 1995 and 2003, scientists were able to perform more than 250 dives with this ROV, discovering 350 new species, including 180 new bacteria, before *Kaiko* was lost at sea during a typhoon when the steel cable connecting it to the mother ship broke.

To be able to withstand high pressures, the cell membranes of the bacteria—which are normally so hydrated that they almost have the properties of a liquid—have to become waxy. Correspondingly, the cell membrane of this

Shewanella strain contains large amounts of omega-3 fatty acids.

Similar setups are also needed for processes inside the cell. High pressure hinders chemical reactions that increase volume, favoring instead the production of relatively rigid, compact molecules. Therefore, the production or even the existence of proteins made of several subunits is very difficult. Nevertheless, such proteins are found in deep-sea bacteria, such as in the ribosomes, which are crucial for all organisms to use their genetic information to produce protein molecules. They always consist of different subunits that have to be combined, and it has been found that the ribosomes of gut bacteria break down into their subunits at pressures above 592 atm (8,700 PSI or 600 bar) in laboratories. How ribosomes can still function in deep-sea bacteria despite these high pressures is completely unknown and is likely to remain so for quite a while, as experiments would have to take place under extremely high pressure.

Other challenges for life in the deep sea are the cold temperatures and the lack of nutrients. Temperature in the deep sea usually ranges from 30.2°F to 39.2°F (-1°C to 4°C), and only about 1 percent of carbon created by biological processes reaches the ocean floor.

Pressure and temperature have a yet-to-be-explained relationship with each other. In labs, when pressure-loving bacteria are cultured at higher temperatures, they need even higher pressures to divide. Conversely, they multiply faster at low temperatures when the pressure is lower than the pressure at which they were found.

S. benthica DB21MT-2 is interesting for several reasons: first, its spectrum of omega-3 fatty acids could be useful as food supplements; and second, the pressure tolerance of its enzymes could have industrial applications, such as carrying out biotech processes under high pressure to increase productivity. Researchers are also studying the bacterium to understand whether and how "normal" bacteria can survive high pressure, because jams, jellies, purees, and juices are sterilized by food manufacturers using high pressure. This procedure doesn't need chemicals, and the color, taste, appearance, and consistency of the foods remain unchanged, as do the nutrients and vitamins. However, *S. benthica* DB21MT-2 proves that bacteria are able to withstand high pressure, so the method might not be as safe as thought.

Janibacter hoylei
Fred Hoyle's
Januslike bacterium

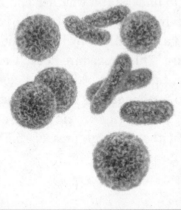

Shape: Spherical or rod
Diameter: 0.4 to 0.7 μm
Locomotion: Nonmotile

IN 2009, INDIAN SCIENTISTS made headlines when they found a previously unknown bacterium in the stratosphere at an altitude of 134,800 feet (41,100 m). It belongs to the genus *Janibacter*, bacteria named after Janus, the two-faced Roman god of beginnings and endings. The bacteria also have two "faces"—they can be spherical or rod-shaped. The scientists named the new bacterium *Janibacter hoylei* after Sir Fred Hoyle (1915–2001), a famous English astronomer.

Hoyle promoted the panspermia hypothesis that life is spread from planet to planet or from solar system to solar system by comets and meteorites. Not only was *J. hoylei* unknown, but it is particularly resistant to the intense ultraviolet radiation at altitudes of 130,000 feet (39,600 m). Could it have found its home in the stratosphere from somewhere beyond Earth?

The stratosphere is considered an extremely inhospitable habitat. Its lower area, at 33,000 to 49,000 feet (10,000–15,000 m) above sea level, is the range where commercial aircraft fly, with temperatures of minus 22°F (–30°C). Rüppell's vulture is thought to hold the record for the highest-flying animal, with a confirmed aircraft impact at 37,000 feet (11,200 m). Higher still, it's slightly warmer, but the atmosphere becomes so thin that birds can no longer breathe. At altitudes of around 130,000 feet (39,500 m), the air pressure is so low that there is almost zero uplift for weather balloons, making it difficult to collect scientific samples at such high altitudes. In October 2014, Alan Eustace ascended to 135,890 feet (41,419 m) before he separated from his high-tech balloon and set the world record for high-altitude free fall. The biggest problem for microbes at these heights, however, is the intensity of ultraviolet radiation, which usually kills them rapidly. Yet, when the scientists found it in the stratosphere, *J. hoylei* was full of life and capable of cell division.

In the meantime, however, the bacterium has also been found down here on Earth. Not only that, but in 2017, an eight-month-old baby in South Korea had a blood infection that was traced to the bacterium. So *J. hoylei* is just as at home on the ground. How did it get into the stratosphere?

Bacteria often travel to the upper atmospheric layers. Hurricanes, tornadoes, volcanic eruptions, forest fires, and so on carry them there on particles of dust or sand. At roughly 33,000 feet, researchers find so many particles with bacteria and other microbes that they believe it's possible for these organisms to influence weather and climate patterns. So-called sprites and jets, "transient luminous events" that shoot upward above thunderclouds, and other phenomena linked to global "electric circuits" can apparently transport these particles even higher. As early as 1978, Russian researchers reported that rockets used to collect high-atmosphere samples had discovered bacteria at 250,000 feet (77,000 m). Forty years later, genetic material from terrestrial and marine bacteria was found in cosmic dust that had collected on the exterior instruments of the International Space Station, which orbits Earth at an altitude of some 250 miles (400 km). It has been experimentally demonstrated that bacteria can survive more than a year under such conditions.

The findings of the Apollo 12 mission of November 1969, however, are still disputed. Astronauts Charles Conrad and Alan Bean landed their lunar module just 600 feet (180 m) away from the unmanned Surveyor 3 probe, which had landed there two and a half years earlier. Conrad and Bean dismantled the probe and brought parts of it back to Earth to analyze the long-term effects of lunar conditions on the materials. The analysis involved testing for bacteria. On the inside of Surveyor 3's camera, scientists discovered *Streptococcus mitis*, bacteria that normally inhabit the human mouth. However, there was doubt about whether the bacteria became established in the camera before or after the trip to the moon. The clean room in which the analyses took place didn't meet today's standards, and the researchers doing the analyses wore short-sleeved shirts instead of the full-body coverings stipulated today.

Halobacterium
salinarum
Salt bacterium

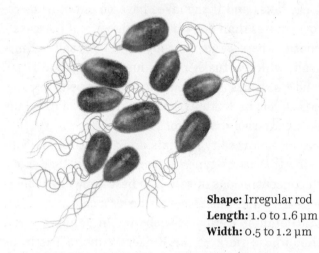

Shape: Irregular rod
Length: 1.0 to 1.6 µm
Width: 0.5 to 1.2 µm

EVERY CHILD KNOWS that seawater tastes salty, but how did the salt get there? What we know as sea salt, table salt, or cooking salt is sodium chloride, a compound of two extremely common and highly reactive elements. In their elemental forms, sodium and chlorine are not found in nature, but they can be produced in a lab where, at room temperature, sodium is a soft whitish metal, and chlorine is a pungent green gas. Because of their reactivity, in nature they only occur bound in minerals, from which water action leaches them as electrically charged ions. Over millions of years, rivers have washed them into the sea, where the sodium chloride content now stands at about 3.5 percent, roughly equivalent to two tablespoons of salt in just over a quart (1 liter) of seawater. A small proportion of sea salt comes from the seafloor and from underwater volcanic vents.

Salt lakes are formed in a similar way. Rivers transport salt to the lakes and if the lakes have no outlet or more water evaporates than flows in, the salt content increases. The deposits where we get our supplies of table salt—Himalayan salt, which actually comes mainly from Pakistan or Poland, as well as salts from Austrian deposits and the Kalahari Desert in South Africa—are residues of dried-up salt lakes, estuaries, or long-gone seas.

There are many salty habitats on Earth, so it's hardly surprising that many types of bacteria are well adapted to high concentrations of salt. The best known, *Halobacterium salinarum*, was discovered over a century ago by German botanist Heinrich Klebahn. In January 1917, Klebahn, whose work at the Kaiser Wilhelm Institute for Agriculture focused on food spoilage, analyzed some spoiled salted cod samples from nearby fish markets. The fish had a reddish coating, which Klebahn scraped off and placed in culture medium. A few weeks later, he noticed red colonies forming; these colonies also grew when the culture medium was saturated with table salt. He named the organism *Bacillus halobius ruber* (the red bacillus living on salt). A few years later, Canadian scientists also discovered red bacteria on salted cod and named them *Pseudomonas salinaria*. Later, it became evident that they were both the same organism. The confusion came to an end in 1930 when microbiologists at their first world congress, in Paris, decided to create an international committee for bacteriological nomenclature. The committee drafted a nomenclature code of practice, which was presented and accepted at the following congress, in London. The now-accepted name of *Halobacterium* was selected at that time with the grammatically incorrect specific name *salinarium*, a mistake that was only corrected in 1996 (the genitive of *salinae* is *salinarum*). Despite its name, *H. salinarum* doesn't belong to the Bacteria domain but to the Archaea.

It grows in salt lakes, salt pans, and salt works and contains a reddish pigment. When *H. salinarum* appears in lakes in great numbers, it gives the lake a red tint. These archaea are eaten by brine shrimps, which in turn are eaten by flamingoes, and the pigment gets into the birds' plumage, giving them their distinctive pink color. *H. salinarum* is also responsible for the production of vitamin A and thus necessary for the maturation of male and female gametes, as well as the growth of flamingo chicks and fledglings. In zoos without ready supplies of brine shrimp, artificial coloring and vitamin A are added to the flamingo food.

H. salinarum can only survive in the presence of salt, and it needs saturated saline solutions with salt concentrations of 20 to 30 percent. If salt content declines too much, the cell wall loses its cohesion and the cell simply dissolves.

Concentrated salt solutions are extreme habitats for other reasons too. They are usually in locations with strong solar radiation, and oxygen is often scarce. Nevertheless, under favorable conditions, a few million salt archaea can live in a single drop of salt water. *H. salinarum* protects itself from light with its red pigment, and it has a highly effective repair mechanism that can quickly mend damage in the genome caused by ultraviolet radiation. In addition, gas-filled vesicles inside the cell refract light. The cell can regulate the amount of gas in the vesicles depending on the light and oxygen content, enabling *H. salinarum* to ascend and descend in salt lakes. It also has an active method of propulsion, using its flagella like a ship's propeller.

Its nutrition is amino acids from decaying organisms in the salt water. Possibly, it also lives in symbiosis with *Dunaliella salina*, a green microalga that also occurs in highly concentrated saline solutions and secretes glycerol. There is evidence that *H. salinarum* not only lives off glycerol but also releases micronutrients that the alga requires. Furthermore, with the aid of purple-pigmented bacteriorhodopsin, it can produce energy from sunlight and in the process changes color to yellow. Three other rhodopsin pigments enable it to orient toward the light or to control salt content inside the cell.

These bacteriorhodopsins, similar to the pigments responsible for vision in the human eye, have launched a whole new field of research: optogenetics. The genes for pigmentation can be selectively implanted into cells. Since the pigments change under the influence of light and trigger a chemical reaction, they can be used as switches to interrupt or restart certain functions inside the cell by using flashes of light at split-second intervals. This has already enabled the research community to investigate the precise sequence of individual steps in chemical processes and the transmission of signals within a cell, and it has even allowed researchers to investigate the transmission of impulses in the nervous system of animals. Biology has a brand-new precision instrument.

H. salinarum moves with flagella; the direction of their rotation changes with certain stimuli. If the living conditions become particularly unfavorable, *H. salinarum* goes into a state of shutdown that can last for thousands of years. Researchers have managed to get microbes that have been trapped inside 10,000-year-old salt crystals to grow again.

Salt never completely dries out. Inside the crystals, there are always minute amounts of trapped water, which are gigantic in comparison with the size of *H. salinarum*. Since other organic matter is usually also trapped there, the archaea have enough nutrients to keep them going, and when supplies eventually run out, they can switch to their shutdown state. Whether they can survive for millions of years, and thus for whole epochs, in salt crystals is a matter of dispute.

Constrictibacter antarcticus

Compact rod from Antarctica

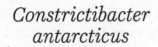
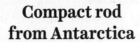
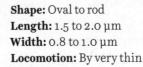

Shape: Oval to rod
Length: 1.5 to 2.0 µm
Width: 0.8 to 1.0 µm
Locomotion: By very thin flagella
Growth form: Often as pairs or chains of cells

CONSTRICTIBACTER ANTARCTICUS was first discovered in 2011 inside a white rock that Japanese scientists had collected a few years earlier in the Skallen region of Lützow-Holm Bay in Antarctica. There, in the east of the icy continent, bare rock protrudes from the ice. The environment is extremely inhospitable—it is bone dry, there is no sunlight during the six-month-long winters, and in the summer the sun shines permanently, flooding the shadeless surroundings with ultraviolet radiation. Except for a few weeks a year, the temperature is below freezing. Still, there are living organisms here. They have had to retreat into rocks to protect themselves from the aridity, the cold, and the ultraviolet light.

Biologists call organisms living in rocks endoliths and divide them into three classes: chasmoendoliths, which live in small cracks and fissures in the rock; euendoliths, which actively bore into rocks, creating tunnels and holes; and cryptoendoliths, which inhabit structural cavities in porous rocks as well as the tunnels and holes created by euendoliths. Endoliths are found in rock samples throughout the world. When these samples come from the Earth's crust in warmer regions, they are often colonized by whole living communities: lichens (organisms somewhere between fungi and microbes that photosynthesize), blue-green algae (cyanobacteria), fungi, bacteria, and archaea, all living with and from one another.

Endoliths can also be found in rocks on or beneath the seafloor and in plutonic rocks deep inside mines or drill cores. As there is no light in their habitat and usually no organic remains from other living organisms, they live a fairly spartan life and are typically chemolithoautotrophs, meaning they produce all the essential molecules and cell components from rock, the inorganic resource of their environment. They produce energy from iron or sulfur, which they extract from rock by secreting acids. They use carbon dioxide or carbonates as sources of carbon.

The extreme lack of energy and nutrients in their environment means that they multiply very slowly. Microbiologists estimate that the cells can divide only once every few hundred years. The record holders are rock bacteria that scientists isolated in 2013 from rock extracted from a depth of 8,000 feet (2,400 m) beneath the ocean floor during the Integrated Ocean Drilling Program.

The 470-foot (143-m) long drilling ship *JOIDES Resolution*, originally contracted by the oil company BP to seek out new oil wells, drilled boreholes in water depths of up to

27,000 feet (8,200 m). On more than a hundred expeditions, the roughly fifty technicians and scientists on board the ship extracted some two thousand seafloor samples and took measurements in the boreholes to learn more about the chemical and physical properties of the oceanic crust. The bacteria discovered in these drill cores, which have not yet been fully characterized, probably only divide once every ten thousand years. Around a hundred million years ago, their habitat was sealed off by the sea, and their numbers are correspondingly small compared with surface soil samples. In deep waters, some ten thousand bacteria can be found in a teaspoonful of rock, while on land in the temperate zones, billions or even trillions are found in a sample of the same size.

Despite the low population density and because the oceanic crust accounts for 60 percent of the Earth's surface, these plutonic rocks beneath the sea are probably the largest ecosystem on the planet, and one that functions completely independently of sunlight. Its natural boundaries are governed by the increase in temperature at increasing depth. At depths of 16,500 feet (5,000 m) below the seafloor, temperatures reach between 300°F and 400°F (150°C–200°C).

Biologists are interested in endoliths like *C. antarcticus* for other reasons. These organisms may be able to exist under conditions such as those found on Mars or some of the moons of Jupiter and Saturn. Such frugal bacteria could probably survive a trip through space in rocks sent spinning off into the darkness by some cosmic catastrophe and then colonize new worlds. So far, however, research on meteorites has failed to produce any clear-cut evidence.

Cupriavidus metallidurans

Heavy metal–tolerant copper lover

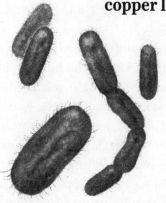

Shape: Rod
Length: 1.2 to 2.2 µm
Width: 0.8 µm
Locomotion: With flagella that cover the whole cell surface
Growth form: Singles, pairs, or short chains

CUPRIAVIDUS METALLIDURANS not only can prosper in environments with high heavy metal content but can also metabolize gold and inspire works of art.

The heavy metal–tolerant, copper-loving bacterium was discovered in 1974 in the wastewater of a zinc smelting plant in the town of Engis in Belgium. The Belgian researchers who discovered it were seeking bacteria that could withstand high metal concentrations, in the hope of identifying microbes that could help restore soils contaminated by heavy metal ions. Even in the Middle Ages, zinc was smelted to produce brass in the vicinity of Engis, and unsurprisingly, the ground is heavily contaminated. The bacterium was later found in other parts of the world, such as Africa, Australia, China, and Japan, also in soils and sediments with high heavy metal concentrations, mostly on industrial sites. But it has also been found in gold nuggets and on the International Space Station. The bacterium tolerates high concentrations of more than twenty different metal ions, including the heavy metals lead, cadmium, chromium, copper, nickel, silver, and zinc, and also cobalt and mercury.

These heavy metals are not only released into the environment from industrial processes or the burning of oil and coal. Major natural sources include volcanic emissions, marine hydrothermal vents, and geysers. Nature is full of metal-containing minerals, and metal-containing habitats are quite common on Earth. Living organisms need metals as trace elements, including those that are toxic at higher concentrations, such as copper, cobalt, zinc, or nickel. Metal ions facilitate the transfer of electrons necessary to generate energy from food. In the human body, half of all enzymes contain metals, and the blood of an average-sized male has around 4.5 grams of iron, the weight of a large nail. Iron is a component of hemoglobin, the pigment in red blood cells that binds oxygen in the lungs and then transports and releases it to the tissues.

Many metal ions become toxic in higher concentrations. This leads to the formation of free radicals, which accumulate in proteins, restricting their flexibility and hindering the active center of enzymes. Heavy metal ions can also displace essential trace elements from their binding sites and thus incapacitate basic cell functions.

C. metallidurans cannot prevent these effects, but it does possess a sophisticated system of proteins to transport metal ions into the cell and back out again. In this way, the bacterium ensures that only the metal ions necessary for survival are present and in the right concentrations.

This all costs energy, but in the environments colonized by this bacterium, there are plenty of supplies, since the high metal pollution means few competitors for food. The genes for the numerous transport proteins are found almost exclusively on mobile genetic elements in the genome, which are known to be rapidly exchanged between different bacteria. The best-known example is antibiotic resistance, which is also spread by mobile elements. Such mechanisms allow bacteria to adapt very quickly to unfavorable conditions and to conquer ecological niches.

The regulatory mechanisms of *C. metallidurans* have become a tool for biotechnology. They can be used like a switch on the genes governing pigment production, making them highly efficient biosensors. With the use of color reactions, they can more or less specifically indicate the presence of certain metals.

If gold salts are present in the environment, *C. metallidurans* can convert them to elemental gold, eventually secreting gold particles. Hence it is actually involved in producing gold nuggets.

To demonstrate this process, in 2012, microbiologist Kazem Kashefi and artist Adam Brown created an installation titled *The Great Work of the Metal Lover*, an allusion to the medieval Magnum Opus, or Great Work, of alchemists searching for ways to turn base metals into gold. For the installation, *C. metallidurans* was in a transparent bioreactor in a colorless gold solution and produced a biofilm of gold particles, which were later isolated and glued onto electron microscope photos of these biofilms.

Magnetospirillum magnetotacticum

Tiny magnetic spiral that orients magnetically

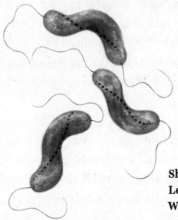

Shape: Crooked spiral
Length: 4.0 to 6.0 µm
Width: 0.2 to 0.4 µm

VISION, HEARING, TOUCH, taste, and smell are all wide-spread senses in the animal kingdom. Whether bacteria can feel, hear, or smell is not known, but they do have the ability to perceive certain physical stimuli and then orient themselves or deliberately move toward or away from objects. They can also react to chemical stimuli. *Magneto-spirillum magnetotacticum* goes one step further: it uses the Earth's magnetic field for orientation. This bacterium was discovered in 1958 by Salvatore Bellini, an Italian physician working at the University of Pavia who was looking for pathogens in water samples. He noticed that certain bacteria on his microscope slide always swam northward and clustered at the northernmost point of the water droplet, and he could divert them with a magnet.

From these observations and after conducting further experiments, he concluded that the bacteria used iron compounds to create a magnetic dipole within themselves, and they used this for orientation.

Bellini's discovery wasn't acknowledged by his peers, because senior scientists at the university considered the discovery implausible and refused to give permission for his two manuscripts to be submitted for publication in a scientific journal. The manuscripts then gathered dust in the archives. This changed in 2007 when they were rediscovered. Until then, the American researcher Richard P. Blakemore was considered to be the discoverer: in 1975 he had observed the same behavior in bacteria from a pond in Woods Hole, Massachusetts. They, too, persistently headed north under his microscope.

Today we know that *M. magnetotacticum* bacteria form structures—so-called magnetosomes—within their cells. Special transport processes pump iron ions from the environment into the magnetosomes, and the chemical environment within these vesicles enables crystals to form. The magnetosomes, in turn, are arranged in chains along the middle of the bacterium's cell axis in such a way that they are not attracted to one another. In this way, they act like a bar magnet or a compass needle. Typically, a cell contains fifteen to twenty of these magnetosomes.

This magnetic sense enables the bacteria to follow magnetic field lines that run into Earth. Presumably, this ability takes them rapidly into the bottom layers of water that they prefer to grow in.

Apart from *Magnetospirillum*, quite a number of other bacteria that form magnetosomes have now been discovered.

Magnetotactic (moving in response to magnetic fields) bacteria like *M. magnetotacticum* have attracted great interest in technology and medicine. Their magnetosomes contain magnetite crystals of a precisely defined size (45 nm). This property is important, because when crystals are too small or too large, they don't have a stable or equally aligned magnetic moment. The magnetosomes formed by bacteria and their magnetite crystals are far superior to artificially produced magnetite particles in terms of shape and consistent size.

Conceivable medical applications might include magnetic resonance imaging, magnetic hyperthermia treatment, and externally controlled delivery of drugs to tumors. In laboratories, bacterial magnetite crystals could be used for unraveling large molecules, for nanosensors and nanoswitches, and so on. Unfortunately, it hasn't been possible yet to culture magnetotactic bacteria in large enough quantities to provide sufficient magnetosomes for commercial use.

The name *Magnetospirillum magnetotacticum*, the "tiny magnetic spiral that orients magnetically," refers in two ways to this bacterium's magnetic properties.

Desulforudis audaxviator

Bold-traveling,
sulfur-reducing rod

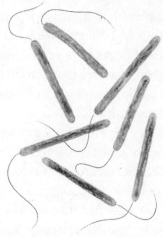

Shape: Rod
Length: 4.8 μm
Width: 0.3 μm
Locomotion: With flagellum
Special characteristics:
Thermophilic; doesn't toler-
ate oxygen

SCIENTISTS HAVE LONG KNOWN that billions of years ago Mars must have been a planet with plenty of water—with rivers and even an ocean. Today, almost 90 percent of the originally available water has disappeared into space. The highly elliptical orbit of the planet around the sun means there are considerable temperature differences between southern and northern hemispheres during the Martian summer. This creates strong winds that transport water vapor to higher altitudes, where the water molecules are split by sunlight. The resulting hydrogen then escapes into space. Today, water is found on Mars only as hoarfrost, which occurs rarely, or in subterranean lakes, one of which was discovered beneath the south pole in 2018.

If there used to be water on Mars, then it is theoretically possible that life arose there too. If so, what has happened to it? The surface is extremely hostile to life, mostly due to ultraviolet radiation from the sun, and water can only be found in the soil. Could it be that life retreated to the depths?

To find out whether this is plausible, scientists from the NASA Astrobiology Institute asked themselves whether there are examples on Earth of organisms retreating to subterranean habitats. Are there bacteria that used to live on the surface but were forced underground millions of years ago and have been there ever since?

In 2006, they found what they were looking for in South Africa's Mponeng gold mine, one of the deepest mines accessible to humans. There, the researchers discovered bacteria, all of one species, in hot water that gushes out of cracks in the rocks almost 1,000 feet (3,000 m) below the surface. They named the bacterium *Desulforudis audaxviator* (officially still *Candidatus*, so its name isn't final).

The zoologist Edward O. Wilson, one of the leading contemporary evolutionary biologists, considered the discovery of a bacterium totally cut off from sunlight in deep rock to be so significant that he immediately made it the subject of his lectures on biodiversity.

It is now clear that *D. audaxviator* had been cut off from life on the Earth's surface for some 25 million years. At that time, the African tectonic plate had just collided with the Eurasian one, and the first apes were colonizing Africa. The climate was temperate, and South Africa was covered by savanna. But the bacterium is not only found in Africa: geomicrobiologists have also found it in deep boreholes and mines in the United States as well as in Europe and Asia. They believe that the bacterium spread through geological processes in deep rock and can be found in all parts of the world.

How it was forced from the Earth's surface to the geologically much older depths has not yet been clarified. *Desulforudis* means "sulfur-reducing rods," with the specific name, *audaxviator*, "bold traveler," a reference to a Latin inscription deciphered by Professor Otto Lidenbrock in Jules Verne's novel *Journey to the Center of the Earth*: *"descende, audax viator, et terrestre centrum attinges,"* descend, bold traveler, and you will reach the center of the Earth. Today the bacterium is perfectly adapted to life deep down: it grows in low oxygen, in complete darkness, and at temperatures of around 140°F (60°C), and it tolerates a fairly alkaline environment with a pH of 9.3. Nutrients, however, are so scarce that researchers believe *D. audaxviator* can multiply at these depths only every one hundred to one thousand years.

The bacterium draws its energy indirectly from the radioactive decay of uranium contained in the deep rocks. The radiation breaks down water molecules, and the bacterium uses the resulting hydrogen to reduce sulfates also found in the rocks. Its nitrogen requirements are covered by ammonium ions from its surroundings, but it also has genes for converting gaseous nitrogen. It uses gases like carbon monoxide or carbon dioxide as a source of carbon, and it can also recycle dead bacteria. It also cannot tolerate oxygen, another indication that it has been isolated from the Earth's surface for a very long time.

The genome of *D. audaxviator* consists of only 2,157 genes, and these genes allow it to recognize and use nutrients, generate energy, and produce all necessary amino acids. It also has a flagellum, enabling it to move. The bacterium can, like a number of other bacteria, produce endospores to survive adverse conditions. At such times, usually triggered by a lack of nutrients, bacteria form a multilayered and almost impenetrable shell, within which the DNA is protected by special proteins. Water content in endospores is greatly reduced, and metabolic activities grind nearly to a standstill. Such endospores can survive for decades and, under particular conditions, even millions of years (*Lysinibacillus sphaericus*; page 63). Once conditions become favorable again, a bacterium can awaken from this prolonged dormancy.

Closely related bacteria have been found in drill cores from great depths and even beneath geothermal springs at depths of 11,500 feet (3,500 m). It is now also known that the bacterium incorporated numerous genes from other bacteria and archaea that it encountered on its evolutionary travels into the depths of Earth.

It's pretty lonely down there: the ecosystem consists of just *D. audaxviator*, rocks, water, gases, and the decaying radioactive elements from which the bacterium gets its energy. On the one hand, it lives there completely independently of sunlight and the rest of the biosphere, since it doesn't need any waste products from other living organisms—not even the oxygen produced by plants. On the other hand, it is the first life-form that lives on the basis of nuclear energy, which is why astrobiologists are very interested in it. They believe that similar organisms could live not only on Mars but on Jupiter's moons Io, Europa, and Ganymede and on Saturn's moons Enceladus and Titan. These places, too, lack oxygen and sunlight but have all the other factors that *D. audaxviator* needs.

Chromulinavorax destructans

Destructive golden algae devourer

(in golden alga, nine hours after infection)

Shape: Spherical
Diameter: 350 to 400 nm (nanometers)

CHROMULINAVORAX DESTRUCTANS is a kind of zombie bacterium, bobbing around as an undead in the freshwater environments of our planet. It shows no signs of metabolic activity, so it neither grows, nor divides, nor multiplies. All this changes when it is swallowed by the golden alga *Spumella elongata*, one of the most common golden algae on Earth that swims in water and lives on bacteria, viruses, and protozoa. Once swallowed, *C. destructans* springs to life and begins eating its victim from the inside out. Microbiologists have aptly named it the destructive golden algae devourer.

It is the alga, not the parasite, that initially does the devouring. *S. elongata* surrounds the bacterium with a small protrusion and guides the bacterium inward to digest its contents. But *C. destructans* blocks this process and takes control of the alga's metabolism. In just three hours, the alga's mitochondria have gathered around the vesicle containing *C. destructans*, and they supply the energy for this zombie bacterium to use the alga's cell components for its own growth and rapid cell division. After twelve hours, two-thirds of the alga is filled with new bacteria, and after another six hours, the alga bursts, releasing the new bacteria into the environment.

Analysis of the genome showed that the bacterium doesn't have a single complete metabolic pathway, meaning that it can't itself produce any of the building blocks necessary to make proteins, genes, carbohydrates, or fats. It can only destroy and remodel already available resources.

This bacterium has a lot in common with a virus—it can only multiply in particular cells and is completely inactive outside of its host. The important difference is that, unlike viruses, *C. destructans* is not dependent on the host's cellular apparatus to replicate its genetic material.

When scientists first discovered this remarkable bacterium, which belongs to an equally strange phylum called Dependentiae (the plural of *dependentia*, Latin for "dependence"), they conducted a systematic search for unknown sequences in RNA libraries and soil samples. The researchers had some success, continued searching, and eventually found the same or highly similar gene sequences in sewage, in hot springs, and in biofilms coating shower heads, water pipes, and sinks. Rather like archaeologists reconstructing texts from fragments of papyrus, the researchers were able to use computers to assemble and evaluate genes and genomes from the sequences. These analyses showed that members of this new phylum had to be parasitic bacteria, that they had comparatively few genes for metabolic enzymes, but that they had many genes for transport proteins, so they apparently lived at the expense of other organisms. Other genes suggested that the host was most likely to be a single-celled eukaryotic organism. At this time, however, the researchers hadn't even found any of these strange bacteria!

This happened once the researchers turned their attention to single-celled organisms that were present in the habitats where the genetic material was found. They determined that the host organisms had to be either amoebas, which use a crawling-like movement, or flagellates, which use whiplike tails for mobility, and that they had to live mainly on a diet of bacteria. And this is how they discovered *C. destructans* as a parasite of the golden alga *S. elongata*.

Bdellovibrio bacteriovorus
Bacteria-eating, leechlike rod-shaped bacterium

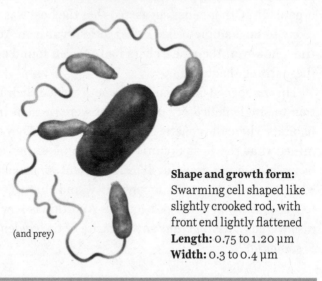

(and prey)

Shape and growth form:
Swarming cell shaped like slightly crooked rod, with front end lightly flattened
Length: 0.75 to 1.20 μm
Width: 0.3 to 0.4 μm

BDELLOVIBRIO BACTERIOVORUS, the bacteria-eating, leechlike rod-shaped bacterium, is the predator among bacteria. In both salt water and fresh water it hunts bacteria like *Escherichia coli* (page 212), *Salmonella*, and *Pseudomonas* (*P. aeruginosa*, page 135) by ramming them, drilling into them, and then feeding on them. Contrary to what the name implies, *B. bacteriovorus* doesn't drain its prey like a leech or tear it to pieces like a shark. It attaches itself to the selected prey with the help of pili—hairlike structures—and uses enzymes to eat a hole through the cell wall. It then enters its prey and closes the hole, ensuring that its host strengthens the cell wall from the inside. This hermetic seal presumably prevents nutrients from escaping while *B. bacteriovorus* begins its destructive work. It remains in the periplasm, the space between the outer and inner membranes of the infected cell. There, it sheds its flagellum and produces enzymes to make the host's inner cell membrane permeable. It then digests the cell's components using transport proteins to convey the nutrients into its own cell interior. The parasite is so thorough that researchers have yet to find any of the host's genes in the genome of *B. bacteriovorus*—nothing remains of the genome of the infected bacterium. While the material within the host cell shrinks more and more, *B. bacteriovorus* grows into a cylinder or tube that eventually divides several times. After a maturation phase, during which new flagella form, the new bacteria dissolve the host's cell wall and exit to search for new victims. The fate of the host cell is quickly sealed—the whole cycle from infection to release takes only three to four hours. Depending on the size of the host cell, three to six and in some cases as many as ninety new *B. bacteriovorus* emerge.

The bacterium records incredible speeds during its hunt: moving at up to 160 µm per second, it can cover more than a hundred times its body length in one second. To be that fast, a person measuring 5 foot 9 inches (1.8 m) would have to move faster than 400 miles per hour (650 km/h). What makes this speed possible for the bacterium is its unusually thick flagellum with kinks. *B. bacteriovorus* doesn't seem to react to chemical stimuli coming from potential victims but rather relies on chance encounters.

Its ability to destroy other bacteria, including many pathogens, makes it interesting for therapeutic purposes.

Experiments on poultry chicks previously infected with *Salmonella* showed that administering *B. bacteriovorus* via the beak led to a significant reduction in infection. It could also be used to combat bacteria in biofilms, bacterial infections of the mouth and throat, intestinal infections, and so on.

The bacteria-eating, leechlike rod-shaped bacterium was discovered by the German microbiologist Heinz Stolp in 1962. It is found in salt, fresh, and brackish waters, in sewage and water pipes, in the soil, on plant roots, and also in the intestines of some animals.

Occupiers of Technological Habitats

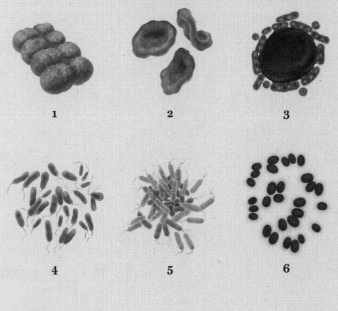

1. *Deinococcus radiodurans*
2. *Dehalococcoides mccartyi*
3. *Alcanivorax borkumensis*
4. *Ideonella sakaiensis*
5. *Burkholderia pseudomallei*
6. *Tersicoccus phoenicis*
7. *Pseudomonas aeruginosa*

Deinococcus radiodurans

Radiation-surviving ball of terror

Diameter: 1.5 to 3.5 µm
Locomotion: Nonmotile
Growth form: Smooth, convex colonies usually of two, four, or eight cells; colored red to pink on culture media

UNDER THE RIGHT CONDITIONS, bacteria divide every twenty minutes. Compare this with the generation time of humans, which for centuries has been roughly twenty years, and it becomes clear that bacterial evolution proceeds at a rapid pace indeed.

It is hardly surprising, then, that bacteria adapt to conditions that we humans have only recently created. Unfortunately, this also includes places that we would like to keep germ-free.

Deinococcus radiodurans, the radiation-surviving ball of terror, is probably the most resistant bacterium in the world. It was discovered when there was still enthusiasm for radioactivity and serious considerations about powering trucks and airplanes with mini nuclear reactors. Radioactivity was used on the hands and faces of clocks and watches to make them glow in the dark—a practice that led to early death for many people working in the watchmaking industry—and it was thought to be useful in preserving food.

In 1956, Arthur W. Anderson had just received his doctorate in microbiology and was working at the Agricultural Research Foundation in Corvallis, Oregon. He began investigating whether canned meat could be preserved using strong gamma radiation instead of heat. He put meat through a grinder and packed it into tin cans, which he then sealed and irradiated with a dose of several dozen grays (Gy)—enough to cause a human being to die within a few days. However, some of the cans began to distend after a few weeks, and on being opened, the tins contained spoiled meat.

The cause was a bacterium that had survived the enormous radiation levels. It made sense to give it the descriptive name of radiation-surviving ball of terror. It also became known by the nickname Conan the Bacterium, after Conan the Barbarian, a fictitious figure from comics and films who dodged almost-certain death on numerous occasions.

Later, the bacterium was found in the cooling water circuits of nuclear reactors. Today we know that *D. radiodurans* can withstand acute radiation exposures of up to 5,000 Gy. At double this dose, only half of the bacteria die, and even with constant radiation of 60 grays per hour, they can live and multiply unimpaired. For comparison: a person can survive exposure to a maximum of 5 Gy, and the intestinal bacterium *Escherichia coli* to 800 Gy. *D. radiodurans* is also extraordinarily resistant to ultraviolet light, which is often used as a disinfectant.

The bacterium protects itself with sophisticated mechanisms: an unusually strong cell wall shields against ultraviolet light, and when damaged, the DNA in the nucleus is patched up in record time by particularly efficient repair enzymes. While normal bacteria can mend two or three damaged spots at once, the "balls of terror" manage to deal with up to five hundred, including the usually fatal double-strand breaks of DNA—a feat of interest to medical research. In addition, the bacterium has several copies of its genetic information and usually occurs in a group of four, with the cells exchanging genetic material with each other that can then be used for repair.

The intriguing question is what led to this extraordinary resilience, because such high levels of radiation do not occur naturally on Earth. One particularly speculative idea was that *D. radiodurans* originally came from Mars, the surface of which is continuously bombarded by strong cosmic radiation. Today it is thought that the bacterium's resistance to radiation is a consequence of its drought tolerance, since *D. radiodurans* also copes particularly well with dehydration.

The bacterium's special abilities are now being used in trials to treat radioactive waste. With certain genetic modifications, it can, for example, make uranium and mercury in nuclear waste insoluble and can break down toxic solvents.

In natural environments, it is found in soils; in excrement and manure; in the human intestine; on meats, dried foods, and medical instruments; in house dust; and on textiles. It can use a wide variety of nutrients and needs oxygen to survive.

Dehalococcoides mccartyi
Perry L. McCarty's
halogen-splitting disk

Shape: Circular with a pronounced indentation on both sides of the disk
Diameter: 0.3 to 1.0 μm
Thickness: 0.1 to 0.2 μm
Locomotion: Nonmotile

ONE OF THE ADMIRABLE PROPERTIES of bacteria is their ability to use almost all complex compounds as sources of nutrition or energy, even those that are extremely dangerous from a human point of view. *Dehalococcoides mccartyi*, for instance, is a bacterium that can break down extremely toxic dioxins, including TCDD, the dioxin that was released when a chemical plant exploded in Seveso, Italy, in 1976. The incident released a total of 6 tons (5.4 metric tons) of various chemicals, among them about 2.2 pounds (1 kg) of TCDD, which settled over 6.9 square miles (18 km²) of the surrounding area. Within a few days, thousands of animals died in the area and several hundred people were affected by skin lesions or chloracne.

The bacterium lives anaerobically in soil and gets its energy by splitting chlorine atoms from dioxins and other related chemical compounds and replacing them with hydrogen. The waste products from this process are easily broken down by other microbes.

Dioxins are organochlorine compounds, molecules predominantly made of carbon and hydrogen in which one or more of the hydrogen atoms are replaced by chlorine. If another element of the halogen group is incorporated instead of chlorine, then it's an organohalogen compound. Most people know chlorine compounds only as artificially produced chemicals (DDT, lindane, PCB, PCP, etc.). In fact, organohalogen compounds are also present in nature: they can result from volcanic eruptions, forest and grassland fires, and geothermal processes. And living organisms can also produce them.

Sponges and other organisms use organohalogen compounds to prevent the growth of bacteria and deter predators. Certain cone snails kill their prey with organobromine compounds; peas, beans, lentils, and other pulses produce organochlorine compounds as hormones; and fungi living in salt water, inland lakes, sewage lagoons, and soils break down nutrients with the help of the enzyme chloroperoxidase and in doing so produce organohalogen compounds in astonishingly high concentrations. Bacteria, in turn, live off these compounds—and *D. mccartyi* is one of them. The natural habitats of these bacteria are anaerobic groundwater, sediments, and sludge in which organochlorine compounds are naturally present. Close relatives can also be found in places far from human activities, such as in the Arctic tundra or in sediment layers beneath the deep sea that are thousands of years old.

As mentioned, dioxins are produced during volcanic eruptions and when organic materials burn above 570°F (300°C) in the presence of chlorine compounds. Situations include forest fires, waste incineration, burning painted or treated wood, and even burning colored candles. The amount and composition of dioxins released depend on the temperature and chlorine content of the combustible materials. More than seventy different dioxins are known and more than 130 furans, which are closely related; they are all collectively known as dioxins. Only seventeen of these compounds are toxic. Since dioxins are very stable, they accumulate in living organisms via the food chain. Humans ingest them mainly through animal foods like fish, meat, eggs, and dairy products.

These compounds gained notoriety after the Seveso chemical plant disaster. Later, substantial amounts of dioxin residues were also found on and near other industrial facilities, such as the former Boehringer Ingelheim chemical plant in Hamburg, Germany, and in the area around Bitterfeld, a major chemical industry center in the former East Germany.

It was in Bitterfeld that *D. mccartyi* was first discovered in 2012, when biologists were looking for bacteria that can tolerate or even break down chlorinated hydrocarbons to remediate chemically contaminated soils.

The bacterium has now been found in many other regions of the world, mostly on contaminated ground. Generally, these sites also have close relatives of *D. mccartyi* that specialize in particular chlorine compounds, as well as other bacteria that further degrade the waste products of *Dehalococcoides* bacteria. Numerous research institutes and companies are now making use of *D. mccartyi* and other *Dehalococcoides* bacteria to break down pollutants like dioxins and other industrially produced organochlorine compounds such as the polychlorinated biphenyls (PCBs), which are plasticizers.

Alcanivorax borkumensis
Alkane eater from Borkum

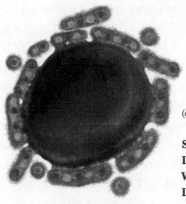

(on a drop of oil)

Shape: Rod
Length: 1.6 to 2.5 µm
Width: 0.6 to 0.8 µm
Locomotion: Nonmotile

ALCANIVORAX BORKUMENSIS LIVES in the sea and was first found in water samples collected near the island of Borkum in the North Sea. It is found only in small numbers in ordinary seawater, but as soon as hydrocarbons—especially the long-chain alkanes—are present, its population explodes. The bacterium is a true specialist, unable to use other nutrients—no sugars, no carbohydrates, no amino acids.

It's not surprising that this bacterium is found in seawater, because even off the main shipping routes, seawater contains hydrocarbons and crude oil from natural sources like underwater asphalt volcanoes and oil deposits. For example, for at least 30,000 years, up to 6,350 gallons (24,000 liters) of crude oil have been seeping out of the seabed daily at the 1.2-square-mile (3-km²) Coal Oil Point seep field off the coast of Santa Barbara, California. Along with the oil come large amounts of methane and other natural gases, all originating from seven extinct asphalt volcanoes. The record-holding seepage site is in the Gulf of Mexico with 175,000 tons (160,000 metric tons) of oil leaking from the seafloor annually. In addition, bitumen and tar sands can be found on the coast and inland. Other hydrocarbons are of organic origin, since numerous organisms produce oils and waxes as protective coatings, as storage for nutrients and fats, and so on.

In places where seawater contains large amounts of oil, *A. borkumensis* proliferates and rapidly becomes the dominant species. But even though it makes a significant contribution to decontaminating oil-polluted seas, coastlines, and beaches, it cannot cope with large amounts of oil, as usually happen with oil spills. It simply can't multiply fast enough for the task. If nitrates and phosphates were added as fertilizers to accelerate the process, the bacteria would quickly use up all the oxygen in the seawater. Furthermore, *A. borkumensis* is only active in surface waters, as it cannot withstand high pressure. Other bacteria take care of oil degradation on the seafloor.

A. borkumensis specializes in exploiting alkanes—long-chained, unbranched hydrocarbons—but can also break down ring-shaped hydrocarbons (aromatics) to a small extent. Crude oil also contains branched alkanes and polycyclic aromatics, which consist of multiple rings and long chains and are feared for their toxicity. Although these substances can also be broken down by bacteria, bacteria are still not a panacea for combating oil spills. However, preparations containing the enzymes of several oil-degrading bacteria could prove useful. The downside of oil-degrading bacteria is that, while breaking down oil, they produce peroxides, acids, and hydrogen sulfide, which in turn are used by other bacteria, thereby lowering the oil's heat resistance and making it more volatile.

The discovery of oil-degrading bacteria in the deep sea is a mystery. At the bottom of the 36,201-foot-deep (11,034-m) Mariana Trench, the deepest known point below sea level, oil-eating bacteria are the dominant bacterial species.

Bacterial density in seawater decreases with increasing depth to about 13,000 feet (4,000 m), and then the numbers gradually increase again. Below a depth of 34,000 feet (10,400 m), they even rise dramatically. Here, oil-degrading bacteria dominate, and in fact, the water at this depth is rich in hydrocarbons that are not identical to those present in the upper layers of the ocean. The origin of these alkanes is so far unknown, but based on the known geology of this region, researchers suspect that they are produced by some yet-to-be-discovered organisms.

Of all the oil-degrading bacteria, *A. borkumensis* is the best studied. It is easy to observe, since the bacteria attach themselves to oil droplets before digesting them. The bacteria transform their cell membranes, making them more lipophilic—having a higher affinity for oil. They develop pili to create better adherence to an oil droplet and form a kind of biofilm around it. In the process, they release surfactants, which act as wetting agents, like those used in dishwashing detergents to dissolve grease. These soap-like substances allow the bacteria to come into contact with and ingest the oil droplets without destroying their cell membrane. They seem to integrate the waste products of the oil into their cell membrane, causing the cells to expand significantly.

Ideonella sakaiensis
Ideon's small
bacterium from Sakai

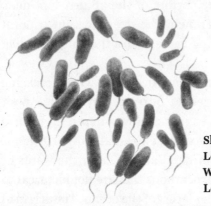

Shape: Rod
Length: 1.2 to 1.5 µm
Width: 0.6 to 0.8 µm
Locomotion: With flagellum

SINCE INDUSTRIAL PRODUCTION and widespread use of polymers such as Styrofoam, polyvinyl chloride (PVC), nylon, polypropylene, and polyethylene terephthalate (PET), plastic waste has become an enormous environmental problem. Most of these products are not biodegradable. That's why there was a small sensation in 2016 when Japanese researchers discovered a bacterium at a recycling plant in the port city of Sakai that was able to break down at least one of these plastics, PET.

This polymer is used in drink bottles and packaging as well as in fibers and textiles, implants, and technical equipment. More than 77 million tons (70 million metric tons) of these plastics were produced in 2020 alone. Experts estimate that it takes 450 years for a PET bottle to decompose.

After reports that yarn made of PET could become moldy, researchers began searching in earnest for organisms that could digest PET. Recycling plants seemed an obvious starting point, since there were large supplies of PET out in the open—a habitat that would likely appeal to PET-degrading bacteria.

Microbiologists filled 250 test containers with a culture medium that contained a piece of PET as the only carbon source and added samples of garbage, wastewater, and soil from the PET recycling plant. Sure enough, communities of bacteria, yeasts, and eukaryotic protozoa were found on the PET. From one of these cultures, the researchers were then able to isolate a bacterium that could break down PET on its own. It grew quickly into a colony, and within six weeks, it had completely decomposed a piece of PET of about one-third of a square inch (2 cm²).

The bacterium from Sakai is closely related to a bacterium that Swedish scientists at the Ideon Research Center had found in activated sludge at a sewage treatment plant in 1994 and named *Ideonella*. That bacterium could digest chlorate found in chlorinated water. The specific name *sakaiensis* comes from Sakai, the discovery site.

I. sakaiensis actually has two enzymes with which it can degrade PET. It attaches to the surface of the plastic object and secretes an enzyme that breaks down PET to an intermediate product called mono-(2-hydroxyethyl) terephthalate (MHET). This so-called PETase is highly specialized and functions at room temperature when the molecular structure of PET is particularly dense. MHET is then taken up by the bacteria and broken down by a second enzyme, MHETase, to terephthalic acid and ethylene glycol, the starting materials for PET production. These chemicals are then used for energy production and are broken down into carbon dioxide and water—a metabolic pathway that other microbes also use.

It is still not certain how these enzymes developed so quickly, because PET has only been around for just over seventy years. Other enzymes like the one that breaks down the herbicide atrazine or certain chemicals used in producing nylon—both being recently developed substances—are known to differ from already existing enzymes by slight mutations. However, enzymes that are closely related to PETase and MHETase have yet to be found.

Genetic engineering has already succeeded in increasing the efficiency of the enzymes and modifying them to break down another intermediate product of PET degradation. This means that PET can be completely recycled and the resulting biomass can be used, or the raw materials used in PET production can be recovered.

If, however, such bacteria were to get into the environment and begin digesting PET waste in the sea, it would be tantamount to fertilizing the oceans—with unknown ecological consequences. On the other hand, it's likely that the same selection process that took place at the recycling plant will, sooner or later, also happen in the huge oceanic plastic-garbage patches formed by winds and currents. Possibly, bacteria that digest microplastics have long been in the sea.

Burkholderia pseudomallei
Walter H. Burkholder's
pseudo mucous pathogen

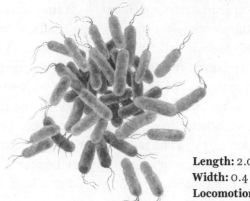

Length: 2.0 to 2.5 µm
Width: 0.4 to 0.8 µm
Locomotion: With flagella

BURKHOLDERIA PSEUDOMALLEI is an expert at coping with hunger. The bacterium is particularly impressive in its ability to survive for decades in distilled water without any nutrients whatsoever. In 1993, in an effort to get to the bottom of such reports, scientists at Mahidol University in Bangkok, Thailand, placed *B. pseudomallei* bacteria in a jar containing nothing but distilled water. The sealed jar was then stored at 77°F (25°C). Once a year, the scientists took a sample to check whether the bacteria were still alive. The result: even after sixteen years, a considerable number of bacteria were present and capable of dividing. Genetic changes in the bacteria suggest that cell division even took place in the jars during this period.

At the same time, *B. pseudomallei* is a dangerous pathogen. Both properties—hardiness and killer potential—make the bacterium attractive as a bioweapon. Such weapons are horrific: silent killers that can be used by terrorists to kill people or incapacitate them for days, weeks, or months without a shot being fired or a bomb dropped. Depending on how infectious it is and its incubation time, a pathogen can spread across continents by trains and airplanes within days and before anyone even realizes that an epidemic has broken out. On the other hand, this is probably the very reason that no state or terrorist organization has used biological weapons in recent times: bacteria and viruses do not distinguish between friend and foe. Every epidemic would eventually reach one's own country and troops. However, this is no protection from crackpots, and bioweapons are easy to obtain. They can be found in nature, just like *B. pseudomallei*, which probably lives on the roots of various plants, where it seems to feed on unicellular organisms like amoebas.

Humans who become infected with this bacterium suffer from melioidosis, an often-misdiagnosed disease that can be fatal, hence the specific name *pseudomallei*. The name *Burkholderia* honors the American plant pathologist Walter H. Burkholder, who discovered the role of many bacteria in plant diseases and their importance in human diseases.

The bacterium is present naturally in the soil in many tropical regions of the world. Infections occur mainly in Southeast Asia and northern Australia, but there have also been cases in Central America, Africa, and Europe. It is estimated that 165,000 people are infected each year, and that half of them die.

Infection occurs through contact with soil or water containing the bacteria, but it's also possible to be infected by inhaling the bacteria, which can cause chronic skin infections and abscesses, pneumonia, and life-threatening sepsis throughout the body. This is why the bacterium is considered a potential bioweapon: it is extremely resistant, can be released in a spray mist, and causes a transmittable disease that is difficult to diagnose and even difficult to treat with antibiotics.

Tersicoccus phoenicis
"Cleanberry"
of the Phoenix probe

Shape: Almost spherical
Diameter: 1 μm
Locomotion: Nonmotile
Special characteristics:
Lives strictly in aerobic
conditions; does not produce
spores

IT IS STILL UNCLEAR whether there is life on other celestial bodies, or at least, signs of prior life. For this reason, it is incredibly important to ensure that spacecraft from Earth landing on or impacting asteroids, moons, or other planets are completely sterile. If they were to introduce organisms from Earth, it could not only falsify the findings from that mission but could have far-reaching ecological consequences for the celestial body. Therefore, spacecraft are manufactured and assembled in certified and constantly monitored ultraclean rooms. As well, all parts and components and all surfaces in the clean rooms are regularly disinfected with ultraviolet radiation and chemicals.

Nevertheless, researchers independently found bacteria in two clean rooms almost 2,500 miles (4,000 km) apart: in 2007 on the floor of a clean room at the Kennedy Space Center in Florida, where the Phoenix spacecraft was being prepared for a Mars mission, and in 2009 in a clean room in French Guiana where the Herschel Space Observatory was reassembled and launched. In both cases, the bacterium in question was *Tersicoccus phoenicis*, which was able to withstand the disinfection methods, the dryness, and above all, the lack of nutrients in the clean rooms. Microbiologists are certain that it also exists in nature, but so far, *T. phoenicis* has not been found anywhere else in the world.

Because it shares less than 95 percent of its genetic information with its nearest relative, it has been classified not only as a new species but as a new genus. Another member of this new genus was discovered in 2016: *Tersicoccus solisilvae*, which—as its specific name suggests—is found in forest soils (the Latin word *silva* means "forest"). It was discovered by Indian scientists studying the microbial biodiversity of the Western Ghats, a mountain range on India's west coast and the country's most biodiverse region.

Tersicoccus the "cleanberry" was so named because of its presence in clean rooms (Latin *tersus* means "clean") and its shape (Latin *coccus* means "large berry"). *T. phoenicis* received its specific name from being found at the Phoenix assembly facility.

In the meantime, scientists have identified nine additional bacterial species in clean rooms and spacecraft, including four close relatives of the extremely resistant bacterium *Paenibacillus xerothermodurans* (page 75). Several hundred more have been isolated and stored in bacterial collections but have yet to be classified.

According to estimates by NASA, the Mars Science Laboratory—a mission that launched in 2011 and landed the rover Curiosity on Mars in August 2012—must have had some thirty thousand spores of heat-resistant bacteria on board. The spores had survived heat sterilization of the structural components.

Even Curiosity itself, a 2,000-pound (900-kg) vehicle the size of a small car, was not bacteria-free. Follow-up analysis of the wipes used to disinfect the spacecraft's surfaces before launching showed that, as well as fungal spores, there was evidence of bacteria from the phylum Actinobacteria, the class Alphaproteobacteria, and the genus *Nitrososphaera*, all of which are present in terrestrial soils and waters.

The findings demonstrate that attempts to rid space probes of all microbes and send them sterile to other celestial bodies are probably futile. It's quite plausible that *T. phoenicis* may be the first living organism from Earth to reach Mars.

Whether Earth bacteria that reached Mars via spacecraft can multiply under the conditions there remains to be seen. Experiments with the common soil bacterium *Bacillus subtilis* showed that perchlorates, which are frequently found in Martian soil samples, as well as iron oxide and hydrogen peroxide, which are also present, seem to be a deadly combination. The addition of ultraviolet radiation, which reaches the Martian surface almost unimpeded, increases the effects: the bacteria died within a minute. On the other hand, simulations show that bacteria may be able to survive in the pores of rocks on Mars.

Pseudomonas aeruginosa
Monas-like, verdigris-colored bacterium

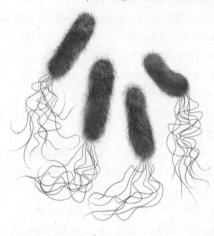

Shape: Rod
Length: 2 to 4 µm
Surface: Covered with adhesive fibers
Locomotion: With multiple flagella attached at one end

THE PERSISTENCE OF SOME BACTERIA can be extremely menacing to humans. It's well known that they can become resistant to antibiotics, but a newer realization is that some organisms, like *Pseudomonas aeruginosa*, have also become resistant to disinfectants—an impending catastrophe for medicine unless counterstrategies can be developed quickly.

P. aeruginosa likes moist and musty conditions. In nature, it is found in swampy regions and damp soils, but it also colonizes humid human-made places: bathrooms, dishwashers, washing machines, dialysis machines, ventilators, catheters, and so on, where the bacteria can be found in hoses. They will also populate the surfaces of toilets, sinks, bathtubs, and seals, forming stubborn and difficult-to-remove biofilms.

From these places, the bacterium can find its way onto and into human bodies and may cause life-threatening infections. It can dissolve red blood cells and secrete toxins. Most often, the people infected have weakened immune systems. The specific name *aeruginosa* (from the Latin *aerugo* meaning "verdigris") alludes to the greenish-blue color of the pus resulting from infection— which can be in the lungs, urinary tract, bloodstream, ear canals, female reproductive organs, intestines, or brain. Why the genus was named *Pseudomonas* in 1894 remains unclear, but possibly the discoverer—Polish German botanist Walter Migula—meant to imply that the shape of these bacteria resembles certain flagellates, which in those days were all termed *Monas*.

The survival skills of this bacterium are impressive. Not only is it involved in producing biosludge in diesel tanks, the so-called diesel plague, but it can also—as long as organic substances are present—grow in distilled water and even in some disinfectants.

Combating it is difficult for several reasons. It clings to surfaces with hairlike appendages called pili, and the outer cell membrane is covered by an alginate that forms a protective sheath. Both the pili and the protective sheath make it difficult to remove these bacteria from the body, for instance from the lungs, and they hinder the effects of antibodies and white blood cells. In addition, the bio-films formed by these bacteria impede the action of anti-biotics as they do disinfectants. Studies have shown that *P. aeruginosa* is three hundred times more resistant to seven common disinfectants when attached to surfaces than when afloat in liquids. Even worse, new research has shown that it not only survives disinfectant treatment, but subsequently becomes resistant to antibiotics to which it was sensitive before the treatment. Furthermore, the bacterium produces certain signal molecules that activate genes in other bacteria. This warning system can initiate the formation of biofilms or the production of protective substances against immune cells.

This example makes it clear that hospitals themselves are now generating new types of pathogens. Today, patients often don't die from infections caught outside hospitals but from "hospital germs"—pathogens that have acquired their virulence and antibiotic resistance in hospitals. Medicine will have to be prepared for the fact that these developments will accelerate in coming years and will have global consequences.

Exotic Eaters

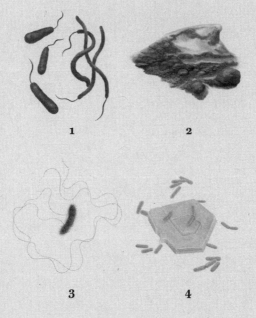

1

2

3

4

Shewanella oneidensis
James Shewan's bacterium
from Oneida Lake

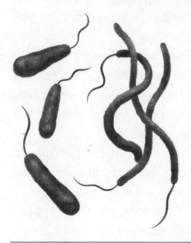

Shape: Extremely flexible and temperature-dependent
At room temperature: Rod, 1.2 to 9.6 µm long, 0.4 to 0.9 µm wide
At temperatures just above freezing: Thread, up to 16 µm long, about 0.5 µm wide
Locomotion: With flagellum at one end

ALL LIFE IS BASED ON THE FLOW of electrons. Humans and animals obtain them from food, use them to produce complex substances, and then transfer them to other molecules, such as oxygen from the air. In this process, called oxidation, the body then exhales carbon dioxide.

The simplest way of acquiring and transferring electrons is from soluble nutrients or gases because they penetrate cells easily. Nutrients and oxygen are scarce in many habitats on Earth, yet bacteria and archaea exist in them. They obtain their electrons from other substances and transfer them to sulfur or nitrogen compounds, carbon dioxide, or metal salts. So, they could be considered as exotic "breathers" of sulfur, nitrate, or arsenic compounds.

Shewanella oneidensis, a widespread soil bacterium, is an example of such a microbe. Its metabolism is not particularly unusual as long as oxygen is available, but if there's no oxygen, *S. oneidensis* can transfer electrons directly to metal compounds. This bacterium belongs to the same genus as *Shewanella benthica* (page 84), and its specific name comes from the site of its discovery in 1988: Oneida Lake in New York State.

S. oneidensis can break down compounds of ten different metals: iron, manganese, palladium, silver, technetium, and vanadium, and also the highly toxic lead, chromium, mercury, and uranium. This is done by certain pigments (cytochromes) in the outer wall of the bacterium that guide the electrons from inside the cell and deliver them to the metal salts. The bacterium can detect the metal compounds in the soil and deliberately head for them. Contact is made with filaments, very thin appendages with a width of 50 to 150 nanometers and a length of tens of micrometers—corresponding to about one hundred times the length of the bacterium's cell. *S. oneidensis* gradually coats metallic minerals with a biofilm in which thousands upon thousands of bacteria grow. There are, however, some compounds that it cannot tolerate, including lithium-nickel-manganese-cobalt oxides, which are used in modern batteries for smartphones, electric bikes, and electric cars.

The bacterium's ability to form elemental silver, mercury, uranium, and so on, makes it interesting for treating soils and wastewaters contaminated by metal salts. It is also well suited to the production of metallic nanoparticles. Early studies have indicated that the size of the particles can be modified by genetically manipulating the surface structure of the bacterium. Such particles could prove useful in treating brain tumors.

Researchers already use the bacterium's ability to transfer electrons to metallic surfaces for manufacturing fuel cells, generating energy or hydrogen from wastewater, and making simple, affordable biobatteries. These batteries consist of folded paper with three surface coatings—one of silver nitrate, a second of conductive plastic, and a third of freeze-dried *S. oneidensis*. The three coated layers are then folded like origami so that the bacterial layer lies between the silver nitrate cathode and the plastic anode. The biobattery is activated by human saliva, which contains enough moisture and nutrients to activate the bacteria. The bacteria deliver electrons from the saliva to the silver nitrate. The battery then produces a current—not enough to charge a smartphone, but enough to run rapid diagnostic medical tests, such as in disaster areas.

The capabilities of *S. oneidensis* go even further. It can produce long, unsaturated hydrocarbons, which it presumably uses as a kind of antifreeze to maintain flexible cell membranes when temperatures are around freezing. If *S. oneidensis* were combined with bacteria that can photosynthesize and produce sugar from carbon dioxide in the air, then reactors could be constructed using both types of bacteria to make fuel from carbon dioxide.

Candidatus Eremiobacter / Candidatus Dormibacter
Sleepers from the desert

(site of the discovery,
Robinson Ridge, Antarctica)

Shape: Rod
Length: Unknown
Width: Unknown
Locomotion: Unknown

COLD, DRY ANTARCTICA is one of the most hostile places on our planet, yet there is life there, such as *Constrictibacter antarcticus* (page 96) in rocks and in the ground. But where do they get the energy needed for their metabolism?

Almost all living communities are based on primary producers. Mostly these producers are plants that use photosynthesis to produce complex organic compounds and that serve as food for other organisms. There are only a few communities in which the primary producers don't get their energy from sunlight. These include the microbes, worms, and snails near black and white smokers in the deep sea. There, the primary producers, such as *Methanopyrus kandleri* (page 72), get their energy from chemical processes. Far away from the smokers, life on the floor of the lightless deep sea is based on sunlight-dependent primary producers, because the remains of dead organisms trickle down from the sunlit layers above.

A few years ago, researchers wondered where the primary producers in Antarctica get their energy—after all, it is pitch dark there for half of the year, and even for the other half, the sun is only just above the horizon for many weeks. Furthermore, there is no flowing water, other than a few seasonal glacial meltwater streams in the Dry Valleys region, where the humidity is so low that neither snow nor ice can collect.

Therefore, scientists took soil samples from Robinson Ridge, a rocky peninsula on the eastern coast of Antarctica, and analyzed them for bacteria. Instead of trying to culture them, they used the modern method of "shotgun sequencing." This process involves isolating the genetic material, multiplying it, and splitting it into fragments at random, as if fired at by a shotgun. The genetic sequence of the fragments is determined using bioinformatics to analyze overlapping sequences and is then reassembled into a complete sequence with as few gaps as possible. With this method, for instance, the words *shotgun sequencing* could provide fragments like shotgu, equenc, unseque, and encing. This operation is child's play for modern computers with plenty of processing power.

Using this method, the researchers were able to reconstruct the genomes of twenty-three microbes. The two most common and not yet described bacteria exhibited an ability that had never been observed before: they use trace gases from the atmosphere to produce carbohydrates from carbon dioxide and water. Algae and plants also produce carbohydrates from carbon dioxide and water, but they need sunlight, which provides the necessary energy to do this through photosynthesis. There are also bacteria like *Clostridium autoethanogenum* (page 179) that can make acetic acid or alcohol from hydrogen and carbon gases if these gases are available in high concentrations. But here it is different. The Antarctic microbes seem to be able to use trace gases with concentrations in the air of only 0.000055 percent by volume (hydrogen) or 0.000025 percent or less (carbon monoxide) to get the energy they need to produce sugar molecules. The carbon source is another trace gas: carbon dioxide, with a concentration in Earth's atmosphere of 0.041 percent by volume. The bacteria have not mastered photosynthesis.

The researchers were able to demonstrate that this process can take place in lab conditions and that the bacteria are also found in another part of Antarctica, Adams Flat. The energy is sufficient to keep the organisms alive even in the Antarctic winters. It is quite possible that bacteria generate energy from trace gases in other ecosystems with scarce supplies of carbon and water.

The bacteria haven't yet been cultured, so we can't fully describe them. However, there are two clearly distinguishable genera that scientists have provisionally named *Candidatus Eremiobacter* and *Candidatus Dormibacter*. The prefix *Eremio* refers to the desert where the organisms live and *Dormi* to the sleep they enter when it's too cold to grow.

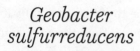

Geobacter
sulfurreducens
Sulfur-reducing
soil bacterium

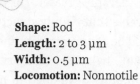

Shape: Rod
Length: 2 to 3 μm
Width: 0.5 μm
Locomotion: Nonmotile
Special characteristics:
Does not produce spores

HUMANS CAN BEND IRON, melt it, forge it, and create alloys with it—but the bacterium *Geobacter sulfurreducens* can breathe it. The sulfur-reducing soil bacterium lives in soils and muddy sediments and breaks down organic material in the absence of air. However, it doesn't ferment its food like other organisms do when oxygen is not available. *G. sulfurreducens* breathes by transferring electrons from its food to the ferric oxide called rust, which is in its environment. Instead of absorbing the iron oxide—the rust particles are usually about half the size of the bacterium—the "iron breather" establishes contact with its pili and uses them to transfer the electrons, like through a wire. These pili—hairlike structures on the exterior of the cell—are made of protein molecules. In

G. sulfurreducens, they are unusually long and can be up to twenty times longer than the bacterium itself.

When American microbiologist Derek Lovley first suggested in 2005 that the bacterium could conduct electrons through its pili like through a metal wire, no one believed him. In biology, electrons are usually transmitted by being passed from one molecule to the next. Metals act differently; electrons are not bound to individual molecules but flow freely, like water in a river, and their conductivity is considerably greater. For most biologists, it seemed inconceivable that simply structured protein molecules like pili could be conductive like metals.

But Lovley saw similarities to synthetic organometallic conductors developed in modern materials science. This field was using aromatic (ring-shaped) molecules of carbon compounds in which the electrons flow freely, as they do in metals.

Lovley and his team provided the proof in 2013 using genetic engineering. They modified the gene responsible for the pili so that normal amino acids were incorporated in place of aromatic ones. The shape of the pili remained the same but without the ring-shaped aromatic components, they could no longer conduct electrons. In addition, the researchers were able to demonstrate in a series of experiments that they could increase or decrease the conductivity of the pili with further genetic modifications and that these changes largely depended on certain amino acids.

Electrically conductive pili (today called e-pili) can now be systematically produced and are therefore building blocks for nanoelectronics based on bacteria.

Experiments by other scientists have also shown that electrons can be relayed to other bacteria, particularly those that store magnetite, such as *Magnetospirillum magnetotacticum* (page 102). There they are stored like in a battery and can be retrieved later. Reliable, effective microbial fuel cells seem conceivable. The amount of electricity produced is small, but it could be enough to supply sensors for monitoring ecological processes in soil or water. Since *G. sulfurreducens* can also break down benzene and other aromatic hydrocarbons, the bacterium could prove useful in treating soils contaminated with oil or production residues.

Ralstonia syzygii
Ericka Ralston's clove bacterium

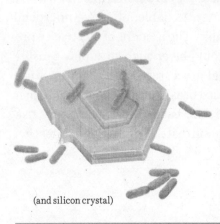

(and silicon crystal)

Shape: Straight rod with rounded ends
Length: 1.0 to 2.5 μm
Width: 0.5 to 0.6 μm
Growth form: Individually, in pairs, and occasionally in short chains
Special characteristics: Does not produce spores

DUST, HAIR, POLLEN, and other particles are the bane of semiconductor production. They contaminate wafers and circuits and reduce yield—a problem that becomes greater as the size of the structures integrated in a circuit become smaller.

This is why semiconductors are produced in clean rooms in which water and air are filtered and personnel wear protective clothing. When producing semiconductors, water is an important raw material used to remove impurities from the surface of wafers after each step of the process. All in all, there are several dozen cleaning phases, so a typical plant producing roughly 40,000 wafers a month uses about 4.8 million gallons (18 million liters) of water a day, with the lion's share being initially processed into ultrapure water. This daily requirement corresponds to the daily water consumption of an American city with a population of more than 50,000 people.

The trouble is that the tanks, pipes, pumps, filters, and spray nozzles for the ultrapure water contain bacteria, as scientists discovered in 2000, when they tried to minimize contamination and found bacteria living in the production lines. They identified the soil bacterium *Ralstonia syzygii* (formerly *Pseudomonas syzygii* and often incorrectly referred to as *R. syzgii*), which had managed to survive all the disinfectant and filtering processes: ultraviolet floodlights and replacement of oxygen and carbon dioxide with pure nitrogen. This alone is not unusual, since other bacteria are known to survive in ultrapure water, such as *Burkholderia pseudomallei* (page 129).

What was surprising, however, was that the bacterium becomes integrated in the wafers. It oxidizes the germanium in the wafer to germanium oxide and produces hydrogen in the process. Thanks to its affinity for the wafer material, it can attach to the wafer surface and also bind materials carried by the rinsing agent. It thus builds a protective sheath, enabling it to withstand further cleaning with aggressive chemicals. Above all, it becomes a flaw to which tiny silicon crystals attach, causing still more damage.

The next surprise was that *R. syzygii* can divide in the semiconductors. It can use trace amounts of wetting agents, alcohols, and other chemicals used in wafer production as food and thereby transfer electrons directly to the semiconductor material. It acts like a transistor.

If a light-sensitive bacterium could be implanted in a semiconductor—or *R. syzygii* suitably modified—then the electron flow could be increased or decreased with light exposure. This could lead to the development of bacterial biotransistors.

R. syzygii is also studied for completely different reasons. In Asia, where cloves are grown, the bacterium causes Sumatra disease in clove trees. It multiplies in the plant's woody vascular tissues, which mainly transport water. The leaves and branches gradually die off until the plant looks scorched, and it dies. *R. syzygii* is transmitted by insects.

The bacterium is named for the American microbiologist Ericka Ralston, who described many new bacteria. The specific name refers to the clove tree, *Syzygium aromaticum*.

Helpful Bacteria

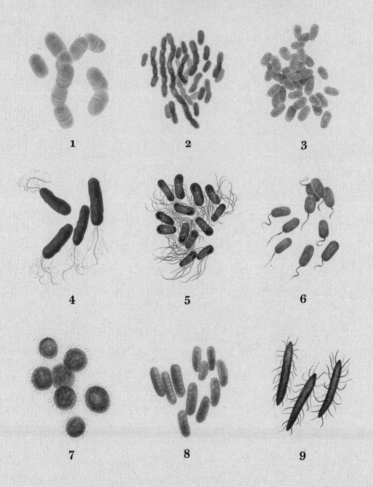

1
2
3

4
5
6

7
8
9

1. *Lactococcus lactis*
2. *Acetobacter aceti*
3. *Propionibacterium freudenreichii*
4. *Bradyrhizobium japonicum*
5. *Bacillus thuringiensis*
6. *Acidithiobacillus ferrooxidans*
7. *Bacillus cohnii*
8. *Cytophaga hutchinsonii*
9. *Clostridium autoethanogenum*

Lactococcus lactis
Milky ball in milk

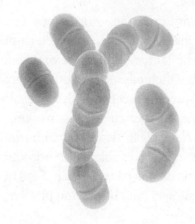

Shape: Round to oval
Diameter: 0.5 to 1.5 µm
Growth form: Mostly in pairs or short chains
Special characteristics: Does not produce spores

ONE OF THE OLDEST bacterial services used by humans is food preservation. Yogurt eaten at breakfast; cheese, salami, sauerkraut, vinegar salad dressings, or soy sauce at lunch; sourdough pastries and black tea in the afternoon; an after-work beer or wine: all are made thanks to bacteria that work alone or alongside yeast. All in all, about one-third of foods eaten today are fermented.

Fermentation has a number of advantages. The foods are safe to eat, last longer, and often taste better. Also, in the course of the fermentation process, toxic or hard-to-digest substances in the natural raw materials are broken down.

Most fermented foods are still produced today using methods that have been around for millennia. Often the processes happened by chance or through inoculation with an already fermented product (old-new inoculation, also called backslopping), so totally undefined cultures are in use that change continuously because strains mutate or disappear due to viral attack.

Lactococcus lactis, the milky ball in milk, is representative of the countless lactic acid bacteria. The bacterium lives in milk, digesting the lactose that milk contains and producing lactic acid, which causes the milk to curdle. Depending on how the process is controlled, the result is butter, buttermilk, kefir, cottage cheese, sour cream, quark, or other products. *L. lactis* is also involved in the initial stages of cheesemaking, as it enables the curds and whey to be separated. Less well known is the role it plays in breadmaking, brewing some specialty beers, and pickling vegetables.

In nature, *L. lactis* is found on plants, mostly grasses, but over the course of more than a thousand years of human use, the lactococci used in production have switched off or lost many genes that wild forms still use.

In medical research, *L. lactis* was the first genetically modified bacterium that was used live for therapeutic purposes. This was for treatment of Crohn's disease, a severe chronic inflammatory bowel disease. Dutch researchers implanted a gene into *L. lactis* bacteria for production of the anti-inflammatory protein interleukin 10 (IL-10). At the end of 2002, the first patients took the bacteria in capsule form. The capsules protected the bacteria from gastric and bile acids and were eventually broken down in the inflamed intestine, releasing the bacteria, which then produced IL-10. Spread of the bacteria in the environment was avoided, because they were missing an essential gene for survival so that they were dependent on supplies of thymine or thymidine. Even though the therapeutic effect was only temporary, the researchers showed that the approach worked and was safe. Several studies are currently underway in which *L. lactis* strains producing other therapeutic proteins are being conveyed to human mucous membranes to treat autoimmune diseases like Crohn's disease, as well as multiple sclerosis, allergies, and arthritis. The advantage is that the human immune system tolerates lactic acid bacteria on mucous membranes because some of these species are part of the natural human bacterial flora.

   L. lactis is also being tested as a probiotic to treat diarrheal diseases or accompany antibiotic treatment, but convincing evidence of effectiveness is currently lacking.

Acetobacter aceti

Acetic acid
rod from vinegar

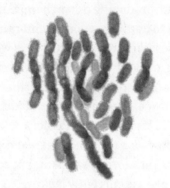

Shape: Straight or slightly crooked rod
Length: 0.9 to 4.2 µm
Width: 0.5 to 0.8 µm
Locomotion: With flagella spread across the cell surface
Growth form: Occurs singly, in pairs, or in chains

ACETIC ACID BACTERIA are found in places where yeast ferments sugars or vegetable carbohydrates to alcohol. They are found on damaged fruits and in the nectar of flowers. They are mostly spread by insects and oxidize fermenting sugar or ethanol via acetaldehyde to acetic acid. Since they fix nitrogen in the air, thus providing plants with usable nitrogen compounds, they can often be found living in symbiosis with some grasses and important crops, including pineapple, banana, mango, coffee, tea, and sugar cane.

People have been using the acetic acid bacterium for at least four thousand years, with archaeological finds providing evidence for the use of vinegar at that time. In Mesopotamia and Egypt, vinegar was made by leaving beer or wine to stand in open containers for months to eventually obtain a sour seasoning or, when diluted with water, a sour drink known in ancient times as *posca*. Together with salted pork and cheese, it was part of the rations for Roman legionnaires on the march. Acetic acid bacteria were carried to the open containers by insects, mostly fruit flies, and gradually turned the alcohol into vinegar.

The bacterium responsible for producing kitchen vinegar, *Acetobacter aceti* or the acetic acid rod from vinegar, was discovered in 1837, but wasn't connected with the formation of vinegar. At the time, this was viewed as a spontaneous chemical reaction. Initially, *A. aceti* was thought to be an alga, and then some type of fungus. Only toward the end of the nineteenth century, with more understanding of bacteria and their differences from algae and yeast, was it realized that *A. aceti* is in fact a bacterium.

In 1864, the French microbiologist and chemist Louis Pasteur was able to prove that it was responsible for the formation of vinegar. Pasteur was convinced about a biological process and analyzed the thin skin that covers vinegar, the so-called mother of vinegar. He managed to isolate an organism from it that produced vinegar in the presence of oxygen. Pasteur's studies had enormous practical implications, because with this knowledge, vinegar makers could speed up production.

This involved adding wood chips, shavings, or the like to alcohol and aerating and circulating the liquid from below. The bacteria, which normally grow only on the surface with access to oxygen, can then, thanks to the circulating air, settle on the pieces of added wood, whose large, rough surfaces enable dense colonization. This speeds up the process considerably, and after a few days, vinegar has formed with up to 12 percent acetic acid. This so-called bondage process—the bacteria are bonded to the carrier material—is only used today for producing high-quality vinegars. Regular kitchen vinegar is produced in fermenters with strong mixing and ventilation in a process that makes vinegar within twenty-four hours.

A. aceti is highly tolerant of acids; the bacterium has very efficient pumps that can pump acetic acid out of the cell. Nevertheless, its cell interior is so acidic that its enzymes need to be protected. How this actually happens is of great interest to industrial microbiology. The bacterium also causes problems in beer and wine production, because it leaves an unpleasant flavor. In the production of bioethanol, it coats the steel components of manufacturing plants with a biofilm and produces acetic acid, thus promoting corrosion.

It is also suspected of being involved in certain rotting processes in damaged fruits. However, it is totally harmless to humans—there is no known disease that is caused by *A. aceti.*

Propionibacterium freudenreichii

Eduard von Freudenreich's propionic acid bacterium

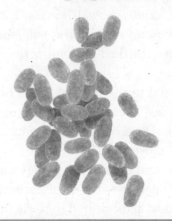

Shape: Oval
Length: About 0.8 μm
Width: 0.4 μm
Special characteristics:
Does not produce spores;
can only survive in
anaerobic habitats

PEOPLE WHO ENJOY EATING Swiss cheese value its nutty, slightly sweet taste and know its characteristic holes. Probably few cheese lovers know, however, that both the taste and the holes are due to the bacterium *Propionibacterium freudenreichii*. Even fewer people are aware that with every bite of Emmental cheese, they are consuming about a billion of these bacteria, which are alive and well in the finished cheese.

Unlike the lactic acid bacteria, Eduard von Freudenreich's propionic acid bacterium ferments lactose not to lactic acid but to propionic acid, hence the name, as well as to acetic acid and carbon dioxide. The latter causes the holes in the cheese. *P. freudenreichii* gets its specific name from the microbiologist Eduard von Freudenreich, who lived in Switzerland from 1851 to 1906. He was the first to systematically investigate the role of bacteria in cheesemaking. He was famous for his textbook *Dairy Bacteriology: A Short Manual for Students in Dairy Schools, Cheese Makers and Farmers* (*Die Bakteriologie in der Milchwirthschaft: Kurzer Grundriss zum Gebrauche für Molkereischüler, Käser und Landwirthe*), first published in 1893 and translated and published in English the following year.

P. freudenreichii is crucial to the taste of Emmental and other Swiss hard cheeses, because it uses the milk fats and metabolic products of other microbes, producing short-chain fatty acids and, above all, esters.

This bacterium is also present in raw milk. Cheesemakers don't only rely on the moderate numbers of naturally occurring bacteria, but instead add cultured bacteria that include not only *P. freudenreichii* but also *Streptococcus thermophilus*, *Lactobacillus helveticus*, and *Lactobacillus delbrueckii*.

P. freudenreichii, however, is the only bacterium that survives the usual processes in Emmental production, such as heating, acidifying, stirring, pouring, seasoning, and maturing at various temperatures. The other bacteria die in the course of cheese production and then become a food source for *P. freudenreichii*.

Since a cheese eater swallows an estimated 30 billion live bacteria with a typical serving of Emmental cheese, the influence of *P. freudenreichii* on human health has been closely investigated. Harmful effects on the intestine or intestinal flora haven't been detected—rather, the opposite: *P. freudenreichii* seems to have a positive effect on the stability of a healthy gut flora, reduce inflammatory processes in the intestine, and possibly have a certain protective effect against colon cancer. It is therefore also found in commercially available probiotic food supplements.

The bacterium is used not only in cheesemaking but also in producing vitamin B_{12}. Genetically modified strains have been made that can produce much more vitamin B_{12} than wild strains of the bacterium. *P. freudenreichii* thus replaces the chemical production of vitamin B_{12}, which requires more than seventy synthetic steps. Vitamin B_{12} is important for blood production and the nervous system and is found naturally mostly in animal food products.

Bradyrhizobium japonicum

Slow-growing Japanese root bacterium

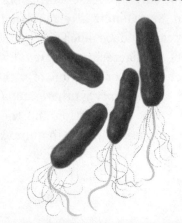

Shape: Rod
Length: 1.2 to 3.0 μm
Width: 0.5 to 0.9 μm
Locomotion: With one thick and several thinner flagella, all arising from the same point at one end of the cell
Special characteristics: Lives in symbiosis with mung beans, cowpeas, and bush beans

BRADYRHIZOBIUM JAPONICUM, the slow-growing Japanese root bacterium, is one of the most economically important bacteria but is probably unknown to most people. The soybean, the basis of millions of people's diet, lives in symbiosis with this bacterium, which supplies the soybean with nitrogen compounds that it extracts from the air. In return, the soybean offers nutrients and protection.

The symbiosis takes place in root nodules with a complex interplay between plant and bacterium and is similar to the actions of other plants in the legume family, including beans, peas, lentils, lupines, and peanuts. They all live in symbiotic relationships, mostly with specific rhizobia (root nodule bacteria). These bacteria also exist outside of plants in the soil and can survive there for decades, although they only multiply slowly.

Nodulation—the formation of root nodules—is initiated by flavonoids that the soy plant releases through its roots. The flavonoids include the pigments responsible for coloring beetroot, red grapes, carrots, apples, and so on. They attract motile root bacteria and ensure that *B. japonicum* produces and secretes a molecule that triggers nodule growth in the plant. It begins with the bacteria anchoring to curved root hairs and forming an infection thread through which *B. japonicum* migrates into the root hair and penetrates the root skin, where it is taken up by plant cells. Eventually, knot-shaped bulges (nodules) are created with cells crammed full of bacteria. Like the bacteria they contain, these bacteroids are not able to divide, but over weeks and months they produce vast amounts of soluble nitrogen compounds. Each nodule can contain millions of bacteria.

Inside the bacteroids, the metabolism of *B. japonicum* changes. It begins to reduce elemental nitrogen from the air to ammonium and excrete it. The enzyme needed for this, nitrogenase, only works well when there is little oxygen. Therefore, the plant regulates the oxygen content in the environment of the bacteroids using a pigment similar to human hemoglobin. The bacteria in turn are supplied with nutrients by the plant. *B. japonicum* has this capacity for fixing nitrogen only within the bacteroid, not in the open.

When the soybean plant dies, the bacteria are released and regain their ability to divide.

The economic significance of root bacteria is enormous. Plants cannot use gaseous nitrogen from the atmosphere; they are dependent on soluble nitrogen compounds like ammonium or nitrates and thus have to be fertilized with mineral fertilizers or nitrogen-rich organic materials to produce high yields. But this is not the case with legumes: they get their supplies of nitrogen from their root bacteria. It is estimated that these bacteria can convert 180 to 270 pounds of atmospheric nitrogen per acre (200 to 300 kg/ha) per year. Globally, they convert between 48 million and 73 million tons (44 and 66 million metric tons) of atmospheric nitrogen and so produce almost half of the nitrogen used as fertilizer in agriculture. Since they produce more nitrogen compounds than the plants can actually absorb, these compounds enrich the soil. This is why legumes are often planted as cover crops to enrich the soil and save on use of mineral fertilizers. Usually, the soil is also inoculated with root bacteria.

If nodule formation and nitrogen fixation could be made more efficient or transferred to other crops, the use of fertilizers could be reduced, and even poor soils could produce high yields. But much more research is needed to reach that point. So far, sixteen genes have been identified as being involved in nodule formation in legumes.

Bacillus thuringiensis
Rod from Thuringia

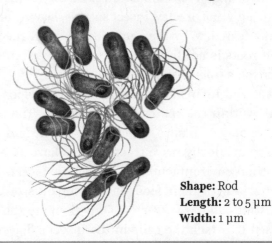

Shape: Rod
Length: 2 to 5 µm
Width: 1 µm

BACILLUS THURINGIENSIS (abbreviated as *Bt*) lives in the soil on and near plant roots. For insects it is a lethal pathogen. It was discovered over 100 years ago independently in both Japan and Thuringia, Germany—in Japan, the microbiologist Shigetane Ishiwatari wanted to find out what was causing the sudden death of valuable silkworms, while in Germany, the microbiologist Ernst Berliner was studying the sudden death of flour moths. Both scientists found the same organism. The Japanese name *Sottokin* (sudden-death bacillus) didn't comply with international nomenclature, and Ishiwatari also didn't clearly identify the bacterium as the cause of silkworm death, so the name *Bacillus thuringiensis* became established. No other bacterium has changed the world of agriculture so much.

There are numerous subspecies of *Bt* bacteria, all of which have one thing in common: they produce toxins each with an effect on very specific insects, including many pests of important crops. Most of the toxins are effective on only one or a few insect species, leaving other species unaffected. Why the bacterium lives in the soil near plant roots is not fully understood, but it seems to protect the roots from pest infestation.

The bacterial toxins are enzymes that are deposited as crystals within the spores of the bacteria. The crystals dissolve only in alkaline conditions. Unlike humans, who have an acidic digestive tract, insects have an alkaline intestinal environment. If a caterpillar ingests the dried bacteria with their food, the crystals dissolve in their intestines and are digested. Only then are the toxins released to attack the cells in the insect's digestive tract. The toxins tear holes in the cell walls, making them permeable to water, salts, and cell components. This causes the rapid demise of cells and the collapse of intestinal function—body fluids and gut contents mix. At first, nerve functions are disrupted, and the insect is paralyzed; a few hours or days later, it dies and is digested by *B. thuringiensis.*

Because of its lethal and, above all, highly selective effects on certain insects, *Bt* has been used as an insecticide since the early 1920s. In 1938, the first *Bt* insecticide containing *Bt* spores and toxin crystals appeared on the market in France under the trade name Sporeine.

B. thuringiensis bacteria are indispensable for organic farming, because they are one of the few effective insecticides approved for organic agriculture. *Bt* is also sprayed over bodies of water to combat mosquito larvae and other blood-sucking insects. However, spraying crops does have drawbacks: the spores and toxin crystals are easily washed away by rain and inactivated by sunlight, and they don't reach insects that eat roots or feed in other inaccessible areas.

Thirty years ago, researchers succeeded in isolating the genes responsible for the production of certain toxins and integrating them into plants like corn, cotton, soybean, and eggplant. These so-called *Bt* plants now produce the toxin proteins in their cells. Because of the toxin's specificity to individual insect pests, it only harms a specific insect species that feeds on the plant, such as the European corn borer or the eggplant borer. For humans and cattle, it is just as harmless as the *Bt* sprays. *Bt* plants have been grown globally for over twenty years and have drastically reduced the use of conventional insecticides. In the United States alone, 7,000 tons (6,400 metric tons) of insecticides are saved annually in growing corn, and there are even greater savings for cotton crops. In Bangladesh, where eggplants are a staple food, farmers can now do without the particularly harmful broad-spectrum insecticides that they previously had to use several times a week and sometimes even twice daily.

Even the side effects of growing *Bt* plants are impressive: neighboring fields with conventional crops benefit from fewer pests, the harvested yield increases, and the impact of molds growing at sites of insect damage drops by 30 to 50 percent. This is also beneficial, as many molds produce liver toxins.

Acidithiobacillus ferrooxidans
Sulfuric acid iron-reducing rod

Length: 1 μm
Width: About 0.5 μm
Locomotion: With flagellum at one end of the cell

THE RIO TINTO, which flows through the Spanish province of Andalusia, owes its red color to iron and copper ions dissolved in the water.

The river has been considered dead for centuries: the water cannot be drunk and no fish live in it. In its upper reaches, there are rich deposits of iron and copper of volcanic origin, and archaeological finds show that these metals were mined here as early as Roman times.

It was probably also the Romans who discovered something peculiar: when solid iron is placed in the strongly sour-tasting water, the iron dissolves and copper is deposited instead. Chemists today know the phenomenon as cementation. That the Romans used this process can be inferred from the remains of water basins and from floors and walls with traces of metallic copper.

Today we know that bacteria are behind the red coloring of the river, the most important one being *Acidithiobacillus ferrooxidans*, the smelting specialist among bacteria. The bacterium is used today in technical processes for metal extraction and soil remediation. In nature, it is found in iron ore deposits, particularly on rock containing pyrite, also known as fool's gold. Chemically, it is iron(II) sulfide.

Pyrite, a gold-colored mineral, can be formed by volcanic processes. In sediments and certain soils, it is formed under the influence of other bacteria that break down organic matter in the absence of oxygen. In the process, they produce hydrogen sulfide and sulfur. Both react with the iron present, resulting in crystalline pyrite over time.

The pyrite, in turn, is used by the sulfuric acid iron-reducing rod bacteria. *A. ferrooxidans* produces iron sulfate and corrosive sulfurous acid as well as sulfuric acid; these are responsible for dissolving other metals from minerals in the area. As the name implies, the bacterium can survive in this self-generated, highly acidic environment at pH of 1 to 2, and it can also be found in the acidic waters seeping out of some mines.

The origin of these highly acidic seepage waters was unclear for a long time. Two American bacteriologists, Kenneth L. Temple and Arthur Russell Colmer, while investigating acid drainage at a coal mine in Montana in 1949, were the first to recognize that the highly acidic waters were of bacterial origin. In 1951, they isolated and described the organism, naming it *Thiobacillus ferrooxidans*; in 2002, the genus *Thiobacillus* was partially reclassified and split based on molecular genetic information. *T. ferrooxidans* was then assigned to the new genus *Acidithiobacillus*.

The discovery had great economic significance. *A. ferrooxidans* was promptly used on an industrial scale for biological leaching, or biomining. This involves making huge piles of crushed rock containing minerals and iron and then spraying them with water. These rock piles can be 3 miles (5 km) long, over a mile (2 km) wide, and more than 300 feet (100 m) high, making them arguably the largest bioreactors in the world. Inside, *A. ferrooxidans* forms biofilms on the rocks along with other bacteria and archaea with similar properties.

The acids produced by the bacteria leach the valuable metals, which are then rinsed out with the water to the foot of the pile. This drainage liquid is spun and rinsed again until the metal concentration is high enough to allow the metals to be separated chemically.

Thanks to the bacteria, metals can even be extracted from low-quality deposits. Roughly a quarter of the copper produced today, more than 5 percent of the gold, and about 3 percent each of the nickel and cobalt are extracted using biotechnology. Uranium is also extracted from ores in this way.

As long as the seepage liquid is properly collected, the only environmental problem is completely separating the acidic wastewater from heavy metals and neutralizing it before disposal. This makes the process considerably more environmentally friendly than the usual smelting methods in blast furnaces.

Bacillus cohnii
Ferdinand Julius Cohn's rod bacterium

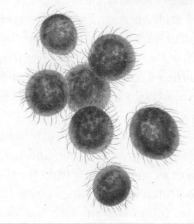

Shape: Round
Diameter: 0.6 to 0.7 µm
Locomotion: With flagella spread all over the cell surface

BACTERIA CAN BE USED not only in the food industry or for metal extraction but in constructing bridges and houses.

Bacillus cohnii, an inconspicuous calcifying bacterium, can do just that. It likes an alkaline environment, like horse manure, which can reach a pH of 8. But it also lives in more alkaline habitats and has now been found all over the world, including in the caustic soda lakes of Europe, Africa, South America, and Turkey, where it produces lime from the dissolved carbonates. It was originally discovered in the early 1990s by bacteriologists at the DSMZ (Deutsche Sammlung von Mikroorganismen und Zellkulturen; German Collection of Microorganisms and Cell Cultures) seeking new bacteria that prefer alkaline conditions. *B. cohnii* was found in a soil sample from a pasture with alkaline soil that contained residues from horse droppings.

B. cohnii not only tolerates alkalies with a pH value over 12 (corresponding to a pungent ammonia solution) but can survive long periods of drought by forming spores. Bacterial spores are extremely resistant and can survive decades or centuries, and under certain conditions even millions of years (*Lysinibacillus sphaericus*, page 63), without losing their ability to germinate. *B. cohnii* is named after Ferdinand Julius Cohn, one of the pioneers of bacteriology in Germany, who, in 1872, was the first to describe the genus *Bacillus*.

The combination of its characteristics—growth in an alkaline environment, ability to produce lime, and long-term survival skills—makes *B. cohnii* interesting for the construction industry. A Dutch microbiologist specializing in lime-producing bacteria came up with the idea of self-repairing concrete. To do this, spores of the bacterium are mixed with ammonium and phosphate salts and a food supply. The mixture is then encased in clay pellets that are less than an eighth of an inch (a few millimeters) wide and these are added to the strongly alkaline concrete. Once it has set, and as long as the concrete remains sealed, nothing happens. If, however, cracks appear over time and moisture seeps in, the spores begin to germinate. The bacteria divide, consume the added nutrients, and produce liquefied calcium carbonate that seals the cracks. Within a couple of days, the bacteria can fill cracks of up to an eighth of an inch (several millimeters).

In this way, *B. cohnii* could solve a huge problem, because cracks in concrete structures regularly necessitate extensive maintenance work and cause damage costing billions. The bacteria could also protect existing buildings. Both sprayed concrete (shotcrete) and liquid concrete repair fluids containing the bacterium have been tested and can be applied to structures with fine cracks.

However, self-repairing concrete is not yet ready for the market. The clay pellets take up too much volume and impair the structure of the concrete. The interaction of the base materials, nutrients, and concrete have to be improved, as do the even distribution and release of the spores, the speed and progression of lime production, and so on. In the meantime, other lime-producing bacteria are being considered for their suitability. However, *B. cohnii* is, so to speak, the pioneer of biological concrete repair.

Calcifying bacteria like *B. cohnii* are now also put to other uses. A German company, for instance, uses them to bind dust created in mining. The bacteria and a liquid culture medium are applied to dusty soil, and within six to forty-eight hours, they produce lime, which binds the dust particles to form sandstone, thus solidifying the dust. Previously, mining companies had to use vast amounts of water to control the buildup of dust. With the help of the bacteria, this water can now be saved.

Cytophaga hutchinsonii

Henry Brougham Hutchinson's cellulose digester

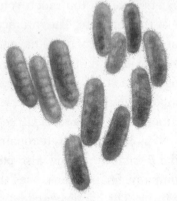

Shape: Flexible rod
Length: 2 to 10 μm
Width: 0.3 to 0.5 μm

CELLULOSE IS PROBABLY the most common organic compound in the world, but surprisingly, there is no animal, apart from a few termites, snails, and mussels, that has enzymes to break down cellulose. This is also surprising, because cellulose, like starch, is made up of glucose subunits. The only difference is the way in which the two molecules are spatially connected to one another. To separate these bonds and be able to break down cellulose into sugar and digest it, animals need microbes. Cows and other ruminants have specially adapted stomachs complete with anaerobic microbes to break down the cellulose. Certain bacteria in the human intestinal flora can also deal with cellulose.

Cellulose molecules are long, unbranched chains of up to tens of thousands of subunits of cellobiose, which in turn consists of two glucose molecules linked together. This makes cellulose a molecule that's difficult to digest even for bacteria; most microbes that digest cellulose do so outside their cell. They secrete large amounts of enzymes, releasing up to 40 percent of their protein content to the outside. This process is "selfless"—since it happens outside the cell, the cellulose digestion enriches the surroundings with large amounts of sugar.

Cytophaga hutchinsonii uses a different tactic. The specific name of this bacterium honors Henry Brougham Hutchinson, who first described the bacterium in 1919 but wrongly classified it as a spirochete, because it had the typical helical shape seen in spirochetes under the microscope. Hutchinson had dried the bacterium to observe it, and the cells twisted in the process, as the Polish microbiologist Helena Krzemieniewska discovered in 1930. A year earlier, the Russian microbiologist Sergei Winogradsky, who was then working at the Pasteur Institute in Paris, had come to a similar conclusion. He suggested the new genus name *Cytophaga* and the specific name *hutchinsonii*.

The special thing about *C. hutchinsonii* is its unique ability to digest cellulose. It forms proteins on its surface that bind to cellulose, and enzymes that cut up cellulose molecules. To do this, they look for regions where the normally solid crystal structure is damaged. The exposed ends are then thought to be threaded through pores or channels into the space between the outer and inner cell walls of the bacterium, where they are further broken down. No sugar is released to the outside—*C. hutchinsonii* is far from selfless, but highly efficient.

The bacterium can quite happily live on damp filter paper, but it does need to be in close contact with its food, on which it glides using a still unknown mechanism. With this gliding mechanism, *C. hutchinsonii* can reach speeds of up to 5 µm per second on glass slides, quite fast for a bacterium, although a common snail moves a hundred times faster. Every now and then, the direction changes abruptly or the bacterium begins to spin rapidly, twice a second. How and why this happens is not yet understood. The cell surface is covered with mobile protein molecules that are arranged like an armored chain or conveyor belt and can contract or relax. In addition, they have tiny fibers that may enable them to attach to surfaces, find suitable locations to begin digesting cellulose, or transport fragments to places where they can be ingested and further digested.

Cellulases—cellulose-digesting enzymes—have been used in industry, agriculture, and sewage treatment plants for decades. Still, there is great economic interest in *C. hutchinsonii*: its efficiency in attacking solid cellulose could help to make the conversion of wood, paper, and plant waste into sugar or alcohol even more efficient. Biofuels would no longer have to be made from starch-based crops.

Clostridium autoethanogenum
Narrow alcohol-producing spindle

Shape: Spindle-shaped rod with pointed ends
Length: 2.1 to 9.1 μm
Width: 0.5 to 0.6 μm
Locomotion: With many flagella
Special characteristics: Anaerobic

FUEL CAN BE PRODUCED not only from the starches of corn specially cultivated as an energy crop or from plant and paper waste, but also directly from exhaust fumes. There are a few microbes that can convert such gases into alcohol. *Clostridium autoethanogenum*, isolated by Belgian researchers from rabbit droppings in 1994, is one of them.

Rabbit droppings are a good source of bacteria that can use carbon oxides, because rabbits have a very specialized digestive system. They have a kind of "shunt" intestine with very little muscle, which is why they constantly have to eat. Newly eaten food shunts the food already being processed in the intestine farther along the digestive tract. Another peculiarity is the cecum, a kind of fermentation sac, which takes up most of a rabbit's digestive tract and is where bacteria break down their food. As the process is not particularly fast, partially digested food is excreted in special feces called cecotropes or "night droppings," which the rabbit then eats. Large amounts of carbon monoxide and carbon dioxide are produced in the process, but these gases cannot escape from the cecum. Other bacteria use the gases, so when the Belgian researchers were searching for new bacterial users of gases, they didn't have to look far, as there were already bacteria known to convert carbon dioxide to acetic acid in the rabbit intestine.

C. autoethanogenum uses a mixture of hydrogen, carbon monoxide, and carbon dioxide as a source of carbon and energy. The hydrogen is used to reduce carbon dioxide. This produces acetic acid and alcohol as well as some by-products. The bacterium lives in the absence of air in the muddy sediments of water. There it is, on the very bottom rung of organic matter decomposition, feasting on plant remains, carcasses, and animal excrement. When the microbes living on these materials and those using the by-products have completed their task, there is no complex organic material left, only gases like hydrogen, carbon monoxide, and carbon dioxide. Only a few archaea and bacteria can make use of these gases to survive.

C. autoethanogenum belongs to the spindle-shaped Clostridia (*Clostridium* means "little spindle") and gets its specific name from its ability to form alcohol (ethanol) independently, without organic carbon.

This species and a few other archaea and bacteria that also use carbon monoxide or carbon dioxide are of interest, because they can be used to produce fuel from industrial emissions. Researchers at a biotech company in Illinois succeeded in identifying the bacterium's metabolic pathways, and by deactivating a gene, they were able to increase the efficiency of alcohol production by 180 percent.

In May 2018, a unit began operating at a steelworks in China, and by mid-2019, the bacteria had generated 9.5 million gallons (36 million liters) of alcohol from its emissions. Production capacity is 19 million gallons (72 million liters) per year. In October 2018, the first transatlantic flight was operated using fuel with its ethanol content produced entirely by bacteria from emission gases of a steelworks. The annual emissions of the iron and steel industry alone, which produces roughly a quarter of all industrial carbon dioxide emissions, could be used to produce about 20 percent of the annual demand for aviation fuel.

There are now similar projects in Germany. To use the carbon dioxide produced from sugar during bioethanol production and thus to improve the carbon footprint of biofuel production, an industrial company is already using bacteria that convert the greenhouse gas into dicarboxylic acids. These are the basic materials needed for the production of plastics like polyester and polyamides (nylon). However, dicarboxylic acids can also be converted into more complex compounds by other microbes in a second step.

The heavily acidic fumes from lignite-fired power plants could also be used to produce basic chemicals. There is great potential here for industry to reduce carbon dioxide emissions.

Dangerous Bacteria

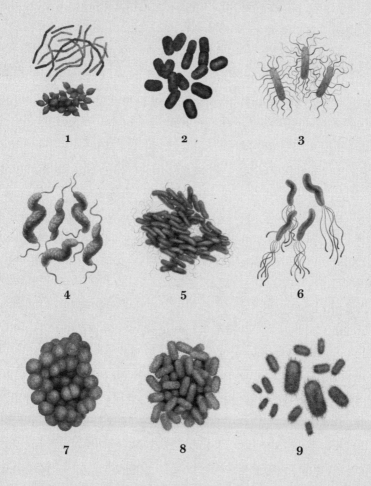

1. *Bacillus anthracis*
2. *Yersinia pestis*
3. *Listeria monocytogenes*
4. *Campylobacter jejuni*
5. *Legionella pneumophila*
6. *Helicobacter pylori*
7. *Staphylococcus aureus*
8. *Serratia marcescens*
9. *Escherichia coli*

Bacillus anthracis
Anthrax-causing rod

(and spores)

Shape: Rod
Length: Up to 10 µm
Width: 1 µm
Spores: Roughly 1 by 2 µm in size
Growth form: Chains of six to eight cells in the blood of infected organisms

BACTERIA COLONIZE NOT ONLY human environments but also people themselves. Some are harmless, others are useful, and still others are fatal.

The soil bacterium *Bacillus anthracis* showed just how close danger can be when envelopes containing *B. anthracis* spores were sent to various senators and news broadcasters in the United States shortly after the 9/11 terrorist attacks. Five people died from receiving those envelopes.

B. anthracis, also known as the anthrax bacterium, is a close relative of *B. thuringiensis* (page 167), so close that scientists are debating whether they are two different species at all. The most distinct differences are in the plasmids of the two bacteria; these are ring-shaped segments of DNA found outside the actual genome of a bacterium and are often exchanged between different species. *B. thuringiensis* has plasmids with genes for *Bt* toxins, while *B. anthracis* has two plasmids with coding for the three subunits of anthrax toxin. Why the respective plasmids are not exchanged between these two bacteria is not known. While *B. thuringiensis* is only dangerous to insects, *B. anthracis* causes anthrax, a serious and often fatal illness that affects grazing animals when they ingest spores of the bacterium through food or water. Since anthrax can be transmitted to humans from animal carcasses, and also through meat, wool, leather, bones, and so on, infections can occur if there is poor hygiene in the handling of animal products or inadequate veterinary monitoring. For centuries, anthrax was a typical occupational disease for people working in ports or in the wool and leather industries. Even today, there are occasional incidents among people in these occupations. In 2010, a 35-year-old folk musician died of anthrax in London after stringing a drum with imported animal hides. There have also been cases of anthrax from infected heroin.

Depending on the route of infection, the disease affects the skin, lungs, or intestines, sometimes with complications such as anthrax meningitis. Since only the spores are infectious, there is no human-to-human transmission. In the early stages of an infection, antibiotics can be used, but if the bacterium has been in the body for a while, the toxins will have already caused irreparable damage. The death rate is very high, particularly with intestinal and pulmonary anthrax, at 40 and 45 percent respectively.

Anthrax spores are taken up by immune cells and germinate there. The released bacteria produce a mixture of three proteins that combine to form toxins and then begin their destructive work.

The largest anthrax epidemic of the twentieth century took place in the Soviet Union in 1979. The cause was an accidental release of spores at the secret military research facility in Sverdlovsk. In contravention of the Biological Weapons Convention negotiated in 1972 and ratified in 1975, the facility had been producing anthrax spores, drying them, and turning them into a powder to be used later in aerosol sprays. For a lung infection, only some three to nine thousand spores need to be inhaled.

On April 2, 1979, an accident while changing an air filter led to roughly one gram of anthrax spores—the equivalent of about fourteen billion spores—escaping from the facility, causing an epidemic in which at least a hundred people died. Government officials blamed contaminated local meat for the anthrax outbreak and denied the real cause for decades.

Anthrax spores are highly resistant to high temperatures, ultraviolet radiation, and aggressive chemicals. They even survive the tanning and preservation of leather and can survive for more than seventy years, possibly even centuries, in the ground. For this reason, the sites of abandoned tanneries are now seen as potentially contaminated with anthrax. For the same reasons, anthrax spores are considered particularly suitable as biological weapons. The bacteria occur naturally in many soils, so they are readily accessible, hence they have been used sporadically in assassination attempts.

Yersinia pestis
Alexandre Yersin's plague bacterium

Shape: Short rod
Length: 1.0 to 1.3 µm
Width: 0.5 to 0.8 µm
Locomotion: Nonmotile
Special characteristics:
Forms a protective sheath
at 98.6°F (37°C)

YERSINIA PESTIS, the plague bacterium, caused the largest known bacterial pandemic in the history of humanity. In the fourteenth century, it killed 30 to 50 percent of the population of Europe. The disease had already killed an unknown number of people in Europe and the Middle East at the time of the Byzantine emperor Justinian, from 541 CE to around 770 CE. Why it returned a few centuries later is not known. The disappearance of the plague in modern times has to do with not only better hygiene but also the displacement of black rats (*Rattus rattus*) by brown rats (*Rattus norvegicus*), which also carry the plague but don't like to be around people.

Y. pestis is named after the Swiss physician Alexandre Yersin, who discovered it while analyzing the swollen glands of victims of the 1894 epidemic that hit Hong Kong. He was working as a ship's doctor and explorer for the Pasteur Institute, and since he was in the South China

Sea when the epidemic started in Manchuria, the French government asked him to go to China to try to discover the cause of the disease.

He was fortunate that the administrators of the British colony hindered his work for political reasons, forcing him to carry out his research at a remote medical station without an incubator. As is now known, the pathogen multiplies only very poorly on culture medium at the usual incubator temperatures, preferring cooler temperatures. Yersin had to culture his agar plates with the bacteria at considerably lower room temperature, and the bacteria grew rapidly. This is why Yersin discovered the pathogen before the Japanese researcher Shibasaburō Kitasato, who had tried in vain to grow his cultures in an incubator, also in Hong Kong.

Y. pestis usually infects rats and other rodents; it is transmitted by fleas, especially the oriental rat flea, *Xenopsylla cheopis*. Once infected, more and more rats die so that eventually fifty to one hundred fleas are living on one rat. When the rat colony is so reduced that fleas have trouble finding a host, they seek other victims, including humans. It takes twenty to twenty-eight days from the first infestation of a rat colony to the first human fatality.

When a flea sucks on the blood of an infected animal, the bacteria get into the insect's intestine and multiply there, forming a clump. Within three to nine days, 300 bacteria grow to some 20,000. This mass of bacteria blocks the flea's gut, so the flea regurgitates the ingested blood, and the bacteria get into the bite wound on the host.

If a person is bitten by an infected flea, the bacteria migrate from the bite area via the lymphatic system to the nearest lymph nodes, where they multiply.

At a body temperature of 98.6°F (37°C), *Y. pestis* forms a protein sheath as protection against an immune response. However, the first bacteria to enter the organism don't yet have this protection. It was only recently discovered that the pathogen forms *Yersinia* outer proteins (YOPs) that are transported into the cells of an infected organism through a kind of hollow needle. This injection structure is only a sixty-millionth of a millimeter in size and protrudes from the bacterium. The YOPs cause the afflicted cells to envelop the pathogens in a protective sheath. *Y. pestis* also produces substances that dampen the immune response and toxins that attack the lungs and liver.

Within a few days, a noticeable swelling, the bubo, begins to form. Other symptoms include headache, body ache, fever, vomiting, and neurological disorders. Death usually occurs from lung or heart failure. The mortality rate is over 50 percent; with antibiotic treatment, the rate drops to 10 to 15 percent. The pneumonic plague, affecting the lungs, is the most serious form and occurs when the pathogen gets into the lungs through airborne droplets. It begins with shortness of breath, blue lips, and coughing up a black, bloody phlegm. This leads to pulmonary edema and circulatory failure. Death comes two to five days later. This form of the plague can be easily transmitted from person to person.

If a lot of bacteria enter the bloodstream, then the course of the disease runs faster, with septic shock, organ failure, and blood clotting. The skin turns a dark purple, which led to the name Black Death. Left untreated, the mortality is 100 percent and those afflicted die on the same day the symptoms first appear.

The bacterium is exceptionally tough. It can live in the ground for up to seven months and on clothing for five to six months. However, it can be deactivated by sunlight.

The plague is by no means gone today. It is endemic in many African countries (Uganda, Madagascar, the Democratic Republic of the Congo), the Americas, and Asia. It isn't spreading anywhere in Europe or Australasia. There are no approved vaccines.

What is particularly worrying is that even in the Middle Ages, the plague was used as a biological weapon by catapulting the corpses of people who had died of the plague over walls and ramparts into cities and fortifications. During the Second World War, Japanese troops experimented with plague bacteria and dropped large numbers of plague-ridden fleas out of planes over Manchuria. Although this only led to minor outbreaks, it spread alarm and terror in the civilian population. And finally, in the Soviet Union in the late 1980s, in violation of the Biological Weapons Convention, genetically modified strains of plague bacteria were produced that were resistant to all available antibiotics.

The "advantages" of *Y. pestis* as a biological weapon are that the disease is highly contagious, the incubation period is short, there is no natural immunity in the population, and diagnosis is difficult. Furthermore, plague bacteria can easily be sprayed with aerosols: just one hundred to five hundred inhaled bacteria are enough to cause pneumonic plague.

On the other hand, the bacterium's uses for humanity include the YOPS and the injection system, which is an effective way of easily transporting anti-inflammatory agents into cells.

Listeria monocytogenes
Joseph Lister's
monocytosis-causing bacterium

Shape: Rod
Length: 1 to 2 µm
Width: 0.4 to 0.5 µm
Locomotion: With numerous flagella over the cell surface
Special characteristics: Does not produce spores

LISTERIA ARE TO BLAME for the fact that unpasteurized milk and raw-milk cheeses less than sixty days matured are banned in many countries. For example, numerous specialty cheeses made in Europe cannot be imported to the United States. Raw milk is considered a health risk, as are the cheeses made from it, cheeses like Brie de Meaux, Camembert de Normandie, Comté, Époisses, Gorgonzola, Reblochon, and Roquefort. In these specialty cheeses, bacteria from unpasteurized milk are responsible for their taste. But unfortunately, these bacteria can include listeria, which cause illness under certain conditions.

Listeria monocytogenes lives on rotting plants and decomposing animal carcasses, but it can also be found in nutrient-poor puddles; in condensation water; in the dust of pillows and cushions; on grasses; in sheep, goat, and cow milk; and in the intestines of humans and animals.

The bacterium can also be found on vegetables cultivated with organic fertilizers. An estimated 10 percent of all people carry the bacteria in their intestine and excrete them in their stool.

L. monocytogenes prefers temperatures of 86°F to 98.6°F (30°C to 37°C) and supplies of oxygen, but it can survive and grow without oxygen and at other temperatures. The lower limit is 39°F (4°C), and the upper 113°F (45°C). This means the bacteria can multiply in chilled, vacuum-packed foods like cold cuts, minced meats, milk, cheese, smoked fish, and ready-to-eat packaged salads.

The genus gets its name from the British surgeon Joseph Lister, who pioneered antiseptic surgery in the 1860s; his ideas and method were initially derided as "Listerism," but now we take them for granted. Lister used phenol to disinfect the operating area and surgical instruments, advocated handwashing and hand disinfection for the surgical team as well as the use of rubber gloves in the operating room, and introduced phenol-soaked bandages for postoperative treatment.

If listeria are ingested in foods, they can cause illness within three to seven days, depending on the dose of bacteria and the person's general state of health. The symptoms for people in good health are nausea, vomiting, and diarrhea—similar to a flu. However, young children and people with weakened immune function can become seriously ill. In these cases, monocytes—a type of white blood cell involved in the body's immune response—multiply rapidly. The bacteria spread in the bloodstream to numerous tissues and organs and can cause sepsis or meningitis. Listeriosis can be treated with antibiotics, but the mortality rate is still 30 percent. While pregnant women hardly notice the infection, their unborn children can die or suffer permanent damage. Infection is possible through the placenta or even during birth. The CDC (Centers for Disease Control and Prevention) estimates that 1,600 people get listeriosis each year in the United States.

L. monocytogenes has developed incredible abilities to overcome various hurdles and survive in human cells. It can penetrate the intestinal mucosa as well as cross the blood-brain barrier and the placental barrier. Once it has entered the bloodstream or the lymphatic fluid, it is either actively absorbed by white blood cells or it penetrates cells using a "zipper" mechanism. In this process, the bacterium induces the host cell to form a protrusion that envelops the bacterium and transports it into the cell.

Once there, the bacterium destroys the protective sac and begins multiplying, doubling itself about once per hour. At the same time, the bacterium binds fiber molecules from the cell, forming a "tail" that can become up to 40 μm long, ten times the length of the bacterium. The steady lengthening of this tail propels the bacterium at speeds reaching 1.4 μm per second. If it comes into contact with the cell membrane of its host cell, another protrusion forms and the bacterium is absorbed by a neighboring cell. There the process is repeated, so that the bacteria are passed from cell to cell without ever being touched by antibodies or white blood cells.

Bacteriophage (a virus that attacks bacteria) preparations are currently being developed using viruses that specifically target listeria. Foods could then be treated with it as a preventive measure.

Campylobacter jejuni
Spiral rod from
the small intestine

Shape: Spiral rod
Length: 0.5 to 5.0 µm
Width: 0.2 to 0.5 µm
Locomotion: With a single flagellum at each end of the cell; the bacterium moves with a corkscrew motion

EVERY YEAR, MILLIONS OF PEOPLE have an unpleasant encounter with *Campylobacter jejuni*. The bacterium—first observed and described by German bacteriologist Theodor Escherich in 1886, though he wasn't able to culture it—is the most common cause of severe diarrhea in humans. Just five hundred bacteria are enough to cause an infection. The incubation time is usually two to five days, and the disease is characterized by watery diarrhea, acute abdominal pain, and fever. Blood in the stool is also possible.

C. jejuni infections occur worldwide, because the bacterium (named for the jejunum, a part of the small intestine) lives in the intestines of a wide variety of animals—wild animals, pets like dogs and cats, livestock, and above all, poultry, which is the most common source of infection. Typically, the surface of the meat becomes infected during the slaughtering process. From there, it gets into food from the hands or kitchen utensils. The bacterium divides fastest at temperatures of around 106°F (41°C), the normal body temperature of birds. Unlike *Salmonella* and *Escherichia coli* bacteria (see page 212), *C. jejuni* cannot multiply outside a host organism. Nevertheless, it can survive for several days in the environment or in food, even withstanding refrigerator temperatures for weeks.

This is why hygiene in the kitchen is so important. Raw poultry meat should be prepared well away from fresh foods, particularly salads. Hygiene experts warn against washing poultry, because bacteria can easily spread to bowls and the sink by splashing water. Infections can also be caused by drinking unpasteurized milk, eating raw or undercooked meats, and being in close contact with pets. Infection from swimming in ponds and lakes is also possible. In some countries, unchlorinated drinking water can be a source of infection as well.

In the intestine, the spiral shape and two flagella make the bacterium highly motile, allowing it to easily penetrate the mucosal wall lining the intestine and attach itself to intestinal cells. While growing, *C. jejuni* releases various toxins that differ from strain to strain. Some destroy intestinal cells, triggering inflammation. The severity of illness depends on the number and type of toxins. People usually get better without treatment, as long as they drink plenty of fluids. In rare cases, antibiotics are needed. For some people, certain surface proteins of the bacterium can cause an autoimmune disease called Guillain-Barré syndrome, which leads to nerve inflammation and symptoms of paralysis. Little is known about the bacterium in poultry, but there is evidence that chickens show mild symptoms when infested with a large number of *C. jejuni* bacteria.

The organization of the genome in *C. jejuni* is peculiar. Unlike most other bacteria analyzed to date, the genes responsible for the elements of certain metabolic pathways are not grouped into functional units, but are spread like a mosaic throughout the whole genome. Also unusual and not yet explained is the almost complete lack of repair systems for damaged genes and the very high variability of some gene sequences.

Legionella pneumophila
Little lung-loving legionnaire

Shape: Rod
Length: 2 to 20 µm
Locomotion: With a flagellum at one end of the cell
Special characteristics: Strictly aerobic

LEGIONELLA PNEUMOPHILA, the little lung-loving legionnaire, gets its name from an epidemic that broke out among members of the American Legion in July 1976. They had all attended a war veterans convention at a large hotel in Philadelphia, Pennsylvania. On returning home, several of the members became ill with pneumonia, as did some other guests and staff at the hotel. Of the 182 people afflicted, twenty-nine died. The unknown disease was then named Legionnaires' disease.

Initially scientists thought a virus was to blame, but six months after the outbreak, they identified a previously unknown bacterium as the pathogen.

L. pneumophila is found in soils, potting soil, compost, and humus and also in fresh water. It leads two different lives. The bacterium can live freely in the environment and forms tough biofilms in water, with up to 650,000 bacteria per square inch (100,000/cm²). Inside this bio-film, the bacteria are protected from harsh living conditions, disinfectants, and antibiotics. At the same time, *L. pneumophila* can live as a parasite inside single-celled protozoa like amoebas and ciliates. It makes contact using pili, and if it's absorbed into the cell, it surrounds itself with a protective covering and releases substances that safeguard its supply of nutrients. Within minutes, the entire metabolism of the affected cell has been hijacked and rearranged, and the bacteria then begin to divide rapidly.

However, the infection doesn't always end in fatality for the host organism. If other bacteria are living as endosymbionts in the host, then *L. pneumophila* multiplies much more slowly and cannot produce certain toxins. Thus it's kept in check.

L. pneumophila can survive drought and starvation inside amoebas and ciliates. In such cases, the unicellular protozoan forms a cyst and goes dormant, only coming back to life when conditions are favorable. *Legionella* can survive in this way for weeks or months.

If the bacteria find their way into water supplies, such as hot water tanks, water pipes, air conditioning units, cooling towers, and so on, they can also form biofilms. When people inhale steam containing bacteria in the shower or in a sauna, *L. pneumophila* infects phagocytes—white blood cells in the bloodstream and tissues—in the lungs and multiplies inside them as it does in amoebas. The results are often fatal, as these phagocytes are actually supposed to absorb and digest bacteria and other foreign materials.

L. pneumophila escapes being digested and multiplies inside the phagocytes, and while doing so, produces toxins and enzymes that damage the lungs and cause severe pneumonia. The elderly, smokers, and people with chronic illness or a weakened immune system are particularly susceptible. The CDC says there were nearly ten thousand reported cases of Legionnaires' disease in the United States in 2018, and the true number may be twice that. Despite antibiotic treatment, the mortality rate is 10 to 15 percent. There is no evidence of person-to-person infection.

As a preventive measure, hot water tanks and piping systems in hotels and swimming pools should be flushed daily with water that's hotter than 140°F (60°C).

Helicobacter pylori
Spiral bacterium
from the stomach

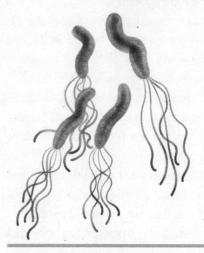

Shape: Spiral rod
Length: 3 µm
Width: 0.5 µm
Locomotion: With four to
six flagella, all coming from
one end of the cell

HELICOBACTER PYLORI LIVES in the human stomach, and it hasn't yet been found anywhere else. Not only has it adapted to the extremely aggressive stomach acids, but it can also cause inflammation, ulcers, and cancer in the stomach and duodenum. This is so unusual that for a long time scientists doubted the existence of the bacterium and its involvement in these diseases.

There was plenty of evidence, though: in Japan, researchers discovered spiral-shaped motile bacteria in the stomach of humans, and in China, researchers found that stomach ulcers could be cured with antibiotics. In Australia, Robin Warren and Barry Marshall observed a strong correlation between stomach ulcers and the presence of this curious bacterium. They suspected a causal link, but other scientists were skeptical, since stress and lifestyle had been considered for decades to be the triggers of gastritis and ulcers. The experts were finally convinced in 1984, when Marshall himself, who was healthy, swallowed a culture of these bacteria taken from the stomach of a patient and then three days later started suffering from gastritis. Twenty-one years later, the two Australian physicians were awarded a Nobel Prize for their research.

Such self-experiments are rare in bacteriology, because they often end fatally. Peruvian medical student Daniel Alcides Carrión died in 1885 after a friend injected him with blood from a wart on a fourteen-year-old boy who had so-called Peruvian warts. Carrión was trying to prove that these warts were a chronic form of Oroya fever. The proof was successful, and today the disease is named after him, but Carrión died and his friend was arrested and charged with murder, though was later released.

Another famous self-experimentation involved the chemist and hygienist Max von Pettenkofer, who disputed Robert Koch's discovery that cholera bacteria were the cause of the disease. In 1892, during a lecture in front of numerous students, he drank a glass of broth containing about a billion cholera pathogens. He developed severe diarrhea but survived the experiment. Presumably, Pettenkofer, then seventy-four years old, had had previous contact with the pathogen and had acquired an immunity that saved his life.

H. pylori is found in the stomachs of people all over the world—more than half of the world's population is infected with it, but only 10 to 20 percent develop inflammation or ulcers in the stomach and duodenum. These can now be treated with a combination of antibiotics. Without treatment or if treatment is delayed, cancer can develop. The pathogen is thought to be transmitted by close contact with people suffering from acute gastroenteritis with vomiting.

Thanks to its flagella, the bacterium can penetrate the thick layer of mucus that coats the stomach wall and protects it from aggressive stomach acids. This is where the bacterium typically stays, attached to the cells of the stomach mucosa. The bacteria establish a neutral zone for themselves by splitting uric acid in the stomach to produce ammonia, which neutralizes hydrochloric acid. If there's a particularly heavy colonization of the bacteria, the ammonia produced can damage the stomach mucosa. For unknown reasons, some types of *H. pylori* produce a protein that penetrates the mucosal cells, changes their shape and size, and ultimately destroys the cells' structure. The results of these processes are inflammation, ulcers, or cancer.

H. pylori has probably been with humans for more than a hundred thousand years. Evidence of its existence was found in the stomach of the five-thousand-year-old glacier mummy Ötzi, the Iceman. Sequences in the genome of *H. pylori* suggest that its ancestors probably lived on a human diet. Comparisons of its genetic fingerprint indicate that the bacteria reached Europe some fifty thousand years ago and that it adapted to the different eating habits of various cultures. For all these reasons, some researchers think that the presence of *H. pylori* in the stomach has a protective function against diarrhea, tuberculosis bacteria, asthma, Crohn's disease, reflux disease, and esophageal cancer. It is, therefore, controversial whether it makes sense to fight this bacterium with a vaccine.

Staphylococcus aureus
Golden bunch of grapes

Shape: Round
Diameter: 0.8 to 1.2 μm
Color: Lab colonies are golden
Locomotion: Nonmotile
Growth form: Often bunched together like grapes
Special characteristics: Does not produce spores

STAPHYLOCOCCUS AUREUS PROVIDES further evidence that seemingly harmless bacteria can very quickly turn deadly. It is found in water and food but also lives on the skin and mucous membranes of humans and animals. Analysis has revealed that almost half of all people are colonized by *S. aureus*, with 20 percent hosting it permanently and 30 percent temporarily. In the course of evolution, the bacterium has adapted very well to humans, so much so that it can colonize people without harming them.

Nevertheless, it can sometimes cause fatal illnesses. The most common and usually least harmful *S. aureus* disease is food poisoning, which is triggered not by an infection but by toxins that the bacterium produces in contaminated foods and that can't be destroyed by cooking. Symptoms of food poisoning from *S. aureus*—especially diarrhea—can begin quickly but may also take a few days to start, and they can last from half an hour to several days.

In humans, *S. aureus* prefers to settle in the nose, where it lives in a community with other bacteria. From there, it can easily spread to a wound with a simple touch of the nose or through sneezing. It can also be transmitted by handshakes, skin contact, shared use of hand and bath towels, and so on.

Once it succeeds in entering a body, it can cause severe infections, though this isn't always the case. On the one hand, susceptibility to infection depends very much on the general state of health of the affected person. On the other, there are at least ten different lineages of strains that affect humans, and they differ significantly in danger. Should the strain be an aggressive variety and the affected person have weak immunity, then *S. aureus* can spread with serious health consequences.

On the skin, *S. aureus* can cause inflammation with boils, abscesses, and swellings. If the bacterium reaches the bloodstream, it can cause inflammation in the lungs, bone marrow, brain, or heart. *S. aureus* is also the most common cause of mastitis. In rare cases, the bacterium can also induce toxic shock or sepsis, which can be fatal.

S. aureus has the ability to remain dormant in the body for a long time. It can penetrate cells and survive there for years until a weakness in the immune system enables an outbreak. Other reservoirs in the body are the surfaces of artificial heart valves, prosthetic joints, and pacemakers, as well as the surfaces of catheters and probes that have been inserted into the body through the skin. The bacterium forms tough biofilms on these devices, and can spread from these surfaces once the conditions are favorable. Within a biofilm, the bacteria are protected from disinfectants.

If an infection occurs, *S. aureus* becomes dangerous through several mechanisms. The bacterium is protected from antibodies and from digestion by white blood cells by a coating and certain proteins on its cell surface. In addition, it manipulates blood clotting to surround itself with a protective wall of fibrin, a coagulated, sticky protein. As soon as it has multiplied, it breaks open the protective wall from the inside and produces enzymes that dissolve the surrounding tissue. The individual *S. aureus* lineages can also produce various toxins that attack white blood cells, perforate cells, and trigger the uncoordinated release of cytokines. Cytokines flooding through the body lead to the very often fatal toxic shock syndrome, in which the circulatory system and organs fail.

S. aureus was discovered in 1881 by the Scottish surgeon Sir Alexander Ogston, and it was much feared in hospitals as the main cause of wound infections. In the 1930s, coagulase testing could identify it with certainty, but still it was untreatable—a dilemma that has been widely discussed in recent times, since genetic tests have been introduced to diagnose untreatable illnesses like Huntington's disease even before it flares up. In 1941, the first patient was cured of infection with *S. aureus* with the newly discovered antibiotic penicillin. Without treatment, about 80 percent of bloodstream infections with *S. aureus* were fatal, but with antibiotics only 15 to 50 percent of patients die, depending on the bacterial strain and the age and general health of the patient. Soon afterward, however, the first penicillin-resistant *S. aureus* strains emerged. They break down the antibiotic using the enzyme beta-lactamase (penicillinase), making it ineffective. As early as 1960, 80 percent of the strains isolated in hospitals were resistant. Today, almost all of them are. In 1959, methicillin was developed, a modified penicillin that can't be destroyed by penicillinase. But in the early 1960s, strains began to appear with methicillin resistance.

Since then, *S. aureus* strains have developed resistance to other antibiotics, usually in the form of multiple drug resistance, and predominantly among methicillin-resistant *S. aureus* (MRSA). Today, the "M" in MRSA stands for *multidrug*. MRSA strains are resistant not only to all beta-lactam antibiotics, but also to a large number of other antibiotics with completely different mechanisms of action. Currently, there are still a few drugs of last resort that are effective against MRSA, but strains of *S. aureus* have been identified that survive the use of some of these antibiotics too. Patients usually receive a course of combined antibiotics that cannot completely eradicate *S. aureus* but can weaken it to the point where the immune system is able to do the rest.

The main sources of MRSA are hospitals, retirement homes, and care facilities, where these bacteria live on devices, surfaces, and all kinds of other objects and are passed from patient to patient by staff. MRSA strains in livestock farming, on the other hand, have practically no role in causing hard-to-treat infections. They are involved in less than 5 percent of cases analyzed and respond well to most therapeutically important classes of drugs. Therefore, 95 percent of the MRSA detected in infections that are hard to treat comes from the field of human medicine.

MRSA shows the need for measures to combat rampant antibiotic resistance. It would make sense to never prescribe antibiotics without first diagnosing the pathogen and its weak point—its sensitivity and resistance to antibiotics. The methods and technologies for rapid diagnosis within a few hours have long been available but haven't been used comprehensively because of the cost. Also, antibiotics shouldn't be available over the counter, which they are in many countries; they should require a prescription everywhere.

Research is desperately needed into new antibiotics as well as other treatment methods. In May 2019, a young girl was cured using phage therapy. She was infected with a life-threatening infection of an antibiotic-resistant bacterium (*Mycobacterium abscessus*) and her chances of survival were slim. The bacteriophages (viruses that attack bacteria) infected the pathogens, reducing them enough that the girl's immune system was then able to eliminate them. Such phages, which only attack certain types of bacteria, are promising as a new approach to treatment.

Serratia marcescens
Serafino Serrati's
decaying bacterium

Shape: Short rod with rounded ends
Length: 0.9 to 2.0 µm
Width: 0.5 to 0.8 µm
Locomotion: With flagella distributed over entire cell surface

SERRATIA MARCESCENS HAS gone down in history as a miraculous bacterium, because, as we now know, it was responsible for the so-called blood miracles that impress many believers year after year—even today. Scientists, however, are more impressed by the versatility and adaptability of the bacterium and by the fact that it is a dangerous pathogen for both humans and coral reefs.

It was discovered in 1819 by the Italian apothecarist Bartolomeo Bizio in Padua when he was trying to trace the reddish discoloration of spoiled polenta. Under the microscope, he discovered a microbe in the red coating and named it *Serratia marcescens*—*Serratia* after the Florentine physicist and steamship designer Serafino Serrati, who had been his physics teacher, and *marcescens* (from the Latin word for "decaying"), because as a colony it quickly becomes a slimy mass.

S. marcescens lives on rotting organic material and can be found practically everywhere: in soil and water, on plants and animals. When it appears in large numbers, it sometimes produces a red pigment. It appeared not only on Bizio's old polenta but also on hosts—wafers of unleavened bread used in many churches for Holy Communion. This is probably the reason behind the blood miracles, in which the hosts appear to bleed, or turn red. This phenomenon has been observed since the introduction of hosts from unleavened bread in the fourteenth century and is viewed by the Catholic faith as transubstantiation, the transformation of bread and wine into the flesh and blood of Jesus Christ during Holy Mass. The red pigment is thus called prodigiosin (from the Latin word *prodigium* meaning "miracle").

The most famous blood miracle took place in Bolsena in 1263 when a Bohemian priest who doubted transubstantiation broke a host during mass and found it red. This miracle prompted the introduction of the Feast of Corpus Christi to the Catholic calendar.

S. marcescens is one of the most resistant and adaptable bacteria. It lives in disinfectants just as happily as in double-distilled water, in contact lens cleaning fluids, and in blood bags. It grows as readily at 39°F (4°C) as at 98.6°F (37°C), survives in the intestines and in seawater, and excretes numerous enzymes that digest proteins, fats, nucleic acids, and insect carapaces. In addition, *S. marcescens* produces serrawettin, a surfactant that allows the bacterium to bond to numerous surfaces and at the same time kills other bacteria.

For a long time, *S. marcescens* was thought to be harmless, and in September 1950, the US Navy even sprayed it over San Francisco Bay during the initially secret Operation Sea-Spray to simulate a biological weapon attack. We now know that it can cause infections of the respiratory and urinary tracts, the eyes, bone marrow, and even the lining of the heart. Most of these infections happen in hospitals during operations or infusions. One particularly susceptible group of people is heroin addicts. Also, cases have been reported from contaminated pharmaceutical products.

Humans aren't the only organisms infected by *S. marcescens*. It also causes white pox disease in hard corals. It has been shown that the bacteria affecting corals originate in human feces that enter the sea in untreated sewage. After infection, corals secrete large amounts of mucus. Within a day, the entire layer of living tissue detaches from the chalky white calcareous skeleton, which is all that remains. How the bacteria manage to adjust from the warm environment of the human intestine to the cool, salty environment of seawater and to an organism with a completely different metabolism remains a mystery. The bacterium can also attack and kill insects: silkworms, grasshoppers, and honeybees. In the case of bees, *S. marcescens* contributes to the death of the colony in winter.

Escherichia coli
Theodor Escherich's
gut bacterium

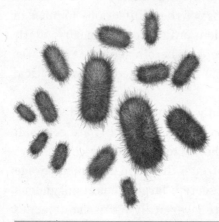

Shape: Cylindrical with domed lid
Length: About 2 µm
Width: About 1 µm
Locomotion: With flagella spread evenly over the cell surface
Special characteristics: Produces acid

IS THERE ANYTHING NEW to say about *Escherichia coli*? Our intestinal inhabitant—known to schoolchildren under the catchy abbreviation *E. coli* or sometimes simply as the coli bacterium—has been the pet of micro- and molecular biologists and the object of bacterial study in medicine and biology for more than a century. And since the human insulin gene was successfully transferred to the bacterium in 1978, it is also the most commonly used production organism for recombinant drugs, vitamins, fine chemicals, and numerous other substances called biologics.

  E. coli can confidently be described as the best-researched bacterium in the world. The internet registers more than a hundred million entries for the bacterium, and the MEDLINE database records more than 380,000 scientific publications related to this organism since 1973. It's hardly surprising that *E. coli* as a study subject has been involved in numerous Nobel Prizes.

The Jekyll-and-Hyde nature of *E. coli* is nevertheless still puzzling: on the one hand, it is a harmless, even necessary intestinal inhabitant of every human being, and on the other, it can turn into a dangerous pathogen.

It has been known since the 1920s that *E. coli* can cause urinary tract infections. Furthermore, it can trigger hospital-acquired infections and sepsis, a blood-borne infection—commonly termed blood poisoning—that can spread to any organ.

It can also lead to intestinal disorders and even cause epidemics with fatal consequences. For instance, in Germany in 2011, more than 4,300 people were infected with an *E. coli* strain that caused not only severe diarrhea but also kidney failure in many of those affected. Fifty people died. The source of the infection was traced to fenugreek sprouts that were sold as a snack or salad ingredient. They were contaminated by *E. coli* strains known as enterohemorrhagic *E. coli*, or EHEC for short. EHEC strains are found naturally in the stomachs of ruminants like cows, sheep, goats, and also deer. Animals that excrete these *E. coli* strains show no signs of disease. The same applies to some people infected with EHEC; they are asymptomatic carriers. The bacteria enter the environment through feces.

EHEC strains can survive for weeks in soil and water. Ingesting a mere ten bacteria may be enough to start an infection. EHEC strains can produce cell toxins that may cause serious diseases. Babies, young children, the elderly, and people with a weakened immune system are particularly vulnerable. For these groups, unpasteurized milk, salami, and raw minced meats like steak tartare are possible sources of infection, as are petting zoos and contact with farm animals for youngsters.

It was recently discovered that *E. coli* strains frequently exchange genetic information with each other, with only a single small section of the genome being transferred at a time. Researchers were able to show more than three thousand examples of metabolic innovations that came about in various *E. coli* strains as a result of such genetic transfer. They enable the respective bacterial strains to survive in an environment that would have been unsuitable if not lethal to its immediate ancestors. The scientists were surprised that the relatively small changes produced such large effects. However, this explains the amazing adaptability of *E. coli*, including changes that turn the harmless gut bacteria into lethal pathogens.

E. coli is named after the pediatrician Theodor Escherich, who discovered it in the intestines (hence the Latin *coli*) of newborns and infants, describing it in 1885 as *Bacterium coli commune*. The following year he published his postdoctoral thesis with the title *Die Darmbakterien des Säuglings und ihre Beziehungen zur Physiologie der Verdauung* (Enterobacteria of infants and their relation to the physiology of digestion), which made him a leading bacteriologist in pediatrics. His whole life was dedicated to hygiene at birth and the care of babies and toddlers. In 1919, eight years after his death, the bacterium was renamed *Escherichia coli* in his honor.

EPILOGUE

WRITING A BOOK ABOUT FIFTY BACTERIA that are important for a variety of reasons was a privilege but also a burden. The greatest struggle was in the omissions.

There are so many other incredible bacteria with astonishing properties that I could have included many more. Then again, there is so much more to explain about each one that my selection could have been smaller.

I had to leave a lot out, because it would lead away from the topic or because too little is known.

The ecology of bacteria is still very much in its early days. It's becoming increasingly clear that bacteria can fundamentally change not only whole ecosystems but even the Earth's atmosphere and climate in relatively short periods of time. Far too little is known about how bacteria interact with other organisms—fungi, higher plants, animals, and above all, humans.

I also could have spent more of the book on communication between bacteria. We know that under certain conditions, bacteria coordinate with each other and they exchange genetic information, but we don't know to what extent this happens and what factors influence these processes. We will learn a lot more about this in the years to come.

Both of these topics—ecology and communication—are of global importance for two reasons. First, bacteria play a far-too-neglected role in climate change. Second, only by understanding their ecology and communication can we ensure that bacterial infectious diseases don't again become a horror for humanity. Only a century ago, a paper cut to the finger could be fatal.

Thanks to the rapid advances in bacteriology, this book deserves a sequel in ten years at the latest. Ten years ago, for example, no one would have thought it possible that bacteria could have something like an immune system. Today, we're astonished by the sophisticated mechanisms these seemingly simple organisms have as protection against hostile viruses. What other hidden abilities have we yet to discover?

Every day, I'm amazed at how complex and diverse these organisms are and what incredible adaptability they show. They have survived cosmic catastrophes, ice ages, warm periods, and several complete transformations of the atmosphere, and they have conquered practically every corner of Earth—life is tough, and it seems that once bacteria gain a foothold in some place, there's no stopping them. This is not only a fascinating thought but an encouraging one.

Last but not least, these tiny organisms are also connected to the many big questions: Why is our planet so alive? Is it the only one in the universe? Is the emergence of living structures extremely unlikely because a large number of factors have to come together simultaneously, or is it inevitable as long as certain basic requirements are met? What is necessary for the further development of such structures in the diversity of living organisms?

If we knew more about the origins of bacteria, we might be able to find the answers to these questions, which ultimately are connected to the question of why we humans exist.

BIBLIOGRAPHY

Berger, Silvia. *Bakterien in Krieg und Frieden: Eine Geschichte der medizinischen Bakteriologie in Deutschland, 1890–1933*. Göttingen, Deutschland: Wallstein Verlag, 2009.

Drews, Gerhart. *Bakterien— ihre Entdeckung und Bedeutung für Natur und Mensch*. Heidelberg, Deutschland: Springer-Verlag, 2015.

Evans, Richard J. *Tod in Hamburg: Stadt, Gesellschaft und Politik in den Cholera-Jahren 1830–1910*. Reinbek, Deutschland: Rowohlt Verlag, 1990. [*Death in Hamburg: Society and Politics in the Cholera Years*. Oxford, UK: Oxford University Press, 1987.]

Gradmann, Christoph. *Krankheit im Labor: Robert Koch und die medizinische Bakteriologie*. Göttingen, Deutschland: Wallstein Verlag, 2005. [*Laboratory Disease: Robert Koch's Medical Bacteriology*, translated by Elborg Forster. Baltimore, MD: The John Hopkins University Press, 2009.]

Perrot, Annick, and Maxime Schwartz. *Robert Koch und Louis Pasteur: Duell zweier Giganten*. Darmstadt, Deutschland: WBG, 2015.

Sarasin, Philipp, Silvia Berger, Marianne Hänseler, and Myriam Spörri, hg [eds]. *Bakteriologie und Moderne— Studien zur Biopolitik des Unsichtbaren 1870–1920*. Frankfurt, Deutschland: Suhrkamp Verlag, 2006.

Yong, Ed. *Winzige Gefährten: Wie Mikroben uns eine umfassende Ansicht vom Leben vermitteln*. München, Deutschland: Kunstmann, 2018. [*I Contain Multitudes: The Microbes Within Us and a Grander View of Life*. London, UK: The Bodley Head, 2016.]

RESOURCES

Micropia, a museum in Amsterdam, the Netherlands, is dedicated to bacteria and other microbes and worth a visit: micropia.nl/en/.

Instructions for DIY manufacturing of Leeuwenhoek microscopes can be found on the website of the Keeling Lab (Department of Botany, University of British Columbia): www3.botany.ubc.ca/keeling/PDF/KeelingMicroscope.pdf.

If you want to do experiments yourself, you will find ideas and equipment at websites like magicalmicrobes.com, learning-center.homescience tools.com, and stevespangler science.com. With kits available from these sites, you can make microbial fuel cells or bacteria lamps.

INDEX

Note: page numbers in italics refer to illustrations.

Halobacterium salinarum;
Janibacter hoylei; Magneto-
spirillum magnetotacticum;
Methanopyrus kandleri;
Paenibacillus xerother-
modurans; Picrophilus tor-
ridus; Shewanella benthica
DB21MT-2

flagella, function, 2, 24, 39
Fliegende Blätter, 8–9
food production, use of bacteria in, 3, 34, 35, 87, 166,
168–69, 194; cheesemaking,
156, 161–63; CRISPR (clustered regularly interspaced
short palindromic repeats),
31, 35; fermentation, 9, 11,
155–56, 158. See also Bacillus thuringiensis
Ford, Brian J., 4–5
Fox, George E., 20–21
Freeman, Victor J., 26
Freeze, Hudson, 19
fungi, 18, 23, 27, 53, 97,
121, 215

genome editing, 31–32
Geobacter sulfurreducens, 139,
147–49, 147
Geogemma barossii
(Strain 121), 74
German Research Founda-
tion, 16
Gram, Hans Christian, 11
Great Oxidation Event, 22
greenhouse gases, 32–33
Guillain-Barré syndrome,
196–97
Gulf of Mexico oil seepage
site, 124

Halobacterium salinarum, 67,
91–95, 91
heat-resistant bacteria, 19, 134,
162. See also Methanopy-
rus kandleri; Paenibacillus
xerothermodurans
heavy metals. See Acidithiobacillus ferrooxidans;
Cupriavidus metallidurans;
Lysinibacillus sphaericus
Helicobacter pylori, 183,
201–3, 201
helpful bacteria. See Acetobacter
aceti; Acidithiobacillus ferrooxidans; Bacillus cohnii;
Bacillus thuringiensis;
Bradyrhizobium japoni-
cum; Clostridium autoeth-
anogenum; Cytophaga
hutchinsonii; Lactococcus
lactis; Propionibacterium
freudenreichii
Herschel Space Observa-
tory, 133
Hesse, Fanny Angelina, 10
Hokkaido, Japan, 79
Hoyle, Sir Fred, 88
Hutchinson, Henry
Brougham, 177

Ideon Research Center, 127
Ideonella sakaiensis, 115,
126–28, 126
industrial uses of bacteria,
34–35, 61, 77, 83, 87,
126–28, 143, 160, 180.
See also Acidithiobacillus
ferrooxidans; Clostrid-
ium autoethanogenum;
Dehalococcoides mccartyi;
enzymes; Minicystis rosea

medicine, bacteria and, 34, 35, 56, 61, 77, 83, 112–14, 142, 156–57. See also *Escherichia coli*; *Magnetospirillum magnetotacticum*; *Pseudomonas aeruginosa*; *Staphylococcus aureus*

melioidosis, 130

metagenomes, 13

metals, bacteria and, 2, 23, 59, 102–3. See also *Acidithiobacillus ferrooxidans*; *Cupriavidus metallidurans*; *Geobacter sulfurreducens*; *Lysinibacillus sphaericus*; *Magnetospirillum magnetotacticum*; *Shewanella oneidensis*

Methanopyrus kandleri, 67, 72–74, 72, 144

Methanosarcina archaea, 28, 29

miasmas, 5–6

microscopes, development of, 4, 9

Migula, Walter, 136

Minicystis rosea, 41, 55–57, 55

molecular biology, 13, 14

Molisch, Hans, x

monocytosis. See *Listeria monocytogenes*

moonmilk, 17

Moszkowski, Alexander: "Überall Bakterien" (Bacteria everywhere), 8–9

Mponeng gold mine, South Africa, 106

MRSA. See antibiotic resistance

Mullis, Kary, 19

Mushroom Spring, Yellowstone Park, 19

Mycoplasma capricolum, 59

Mycoplasma genitalium, 50, 59

Mycoplasma laboratorium, 58–59

Mycoplasma mycoides, 59–60

myxobacteria, 55–57

Nakhla meteorite, 64

nanoelectronics, 148

NASA Astrobiology Institute, 106

Nasu, Socho, 54

Nasuia deltocephalinicola, 41, 52–54, 52

Nitrososphaera (genus), 134

Ogston, Alexander, 206

oil-degrading bacteria. See *Alcanivorax borkumensis*

Oneida Lake, 142

Operation Sea-Spray, 210–11

optogenetics, 94

organohalogen, 121

Oroya fever, 202

Ötzi, the Iceman, 203

ox warble fly, 83

Paenibacillus (genus), 76, 77

Paenibacillus xerothermodurans, 67, 75–77, 75, 133

panspermia hypothesis, x, 64, 88

paramecia, 18

Paramecium, ix

Pasteur, Louis, 6–7, 159

Pelagibacter ubique, 41, 49–51, 49, 55, 80

Permian–Triassic extinction event, 28–29

Petr Kottsov (Russian research ship), 44

petri dishes, ix, x, 10

Petri, Julius, 10

phagocytes, 199–200